U.S. Marine Corps Combat Water Survival

Plus

Marine Combat Stress

Department of Defense

Pentagon
Publishing

MCRP 3-02C

Marine Combat Water Survival

U.S. Marine Corps

PCN 144 000069 00

To Our Readers

Changes: Readers of this publication are encouraged to submit suggestions and changes that will improve it. Recommendations may be sent directly to Commanding General, Marine Corps Combat Development Command, Doctrine Division (C 42), 3300 Russell Road, Suite 318A, Quantico, VA 22134-5021 or by fax to 703-784-2917 (DSN 278-2917) or by E-mail to **morgann@mccdc.usmc.mil**. Recommendations should include the following information:

- Location of change
  - Publication number and title
  - Current page number
  - Paragraph number (if applicable)
  - Line number
  - Figure or table number (if applicable)
- Nature of change
  - Add, delete
  - Proposed new text, preferably double-spaced and typewritten
- Justification and/or source of change

Additional copies: A printed copy of this publication may be obtained from Marine Corps Logistics Base, Albany, GA 31704-5001, by following the instructions in MCBul 5600, *Marine Corps Doctrinal Publications Status.* An electronic copy may be obtained from the Doctrine Division, MCCDC, world wide web home page which is found at the following universal reference locator: **http://www.doctrine.usmc.mil**.

Unless otherwise stated, whenever the masculine gender is used, both men and women are included.

DEPARTMENT OF THE NAVY
Headquarters United States Marine Corps
Washington, D.C. 20380-1775

6 January 2003

FOREWORD

Marine Corps Reference Publication (MCRP) 3-02C, *Marine Combat Water Survival*, provides Marine Corps combat water survival techniques, procedures, and training standards. This publication also teaches Marines to cross water obstacles and perform water rescues correctly and safely.

This publication is the foundation for teaching Marines correct water survival techniques and procedures that are used throughout the Marine combat water survival program (MCWSP). Once an individual or a unit has completed the MCWSP, this publication can be used as a refresher course before water operations.

MCRP 3-02C supersedes Fleet Marine Force Reference Publication (FMFRP) 0-13, *Marine Combat Water Survival*, dated 16 September 1991.

Reviewed and approved this date.

BY DIRECTION OF THE COMMANDANT OF THE MARINE CORPS

EDWARD HANLON, JR.
Lieutenant General, U.S. Marine Corps
Commanding General
Marine Corps Combat Development Command

Publication Control Number: 144 000069 00

MCRP 3-02C Marine Combat Water Survival

Table of Contents

Chapter 1. Survival at Sea

Chapter 2. Water Rescues

Chapter 3. Treatment of Casualties and Avoidance of Dangerous Marine Life

Chapter 4. Negotiating Water Obstacles

Chapter 5. Fording Waterways

Appendix. Knot Tying

Throughout history, water has posed special challenges to Marines and Sailors during times of both peace and war. Therefore, the inherent dangers associated with waterborne operations demand that Marines and Sailors receive proper water survival training. Combat units that are confident in their ability to work in and around water can use the water to their advantage in combat, and history is filled with examples of how the proper preparation or training for survival in water reduced or averted disaster.

USS *Indianapolis*

On Sunday, 29 July 29 1945, the heavy cruiser USS *Indianapolis* (the *Indy*) was en route to the Philippine Sea. Shortly before midnight (about 39 hours out of port), the *Indy*, running blacked out and unescorted, was rocked by two explosions on her starboard side. With communications smashed, the ship could not signal its distress and sank within 15 minutes.

Three life rafts and a floater net supported a few survivors, but the rest of the survivors drifted about, held up by rubber life belts or Mae Wests. About 60 seamen died the first night. Survivors assumed the ship would be reported overdue in Leyte, and that they would be rescued within 2 days. By Monday evening, however, panic began to set in as some life jackets lost their buoyancy from the long immersion. Some men even fought over life jackets, which resulted in at least 25 deaths. No one dared sleep for fear of losing his jacket.

Throughout the next several days, in-transit aircraft flew nearby without spotting the desperate seamen. As best they could, the men kept together, some tying long ropes to each other, floating like corks on a net.

Not until late Thursday morning, 3 1/2 days after the ship sank, were the men discovered. Luckily, a plane on a routine flight over the area spotted the survivors. When ships picked them up that night, the survivors learned they had never been reported overdue. Every one of the 1,196-man crew was a casualty; 880 crewmembers were listed as dead or missing. Although many lives were lost, the innovative and expedient use of flotation devices and float techniques employed by the survivors helped save hundreds of lives.

USS *America*

On Thanksgiving eve, 23 November 1995, the USS *America* made its way through the Arabian Sea. Twenty-year-old, Marine Lance Corporal (LCpl) Zachary Mayo was unable to sleep and, wanting some fresh air, made his way onto an open-air platform near the aircraft hangar bay, which was three levels below the sleeping quarters. While he was on the platform, the ship veered suddenly, throwing LCpl Mayo through the platform's protective bars and into the sea, 30 feet below.

Frantic, LCpl Mayo called out in vain to the watchmen on the flight deck, which was 64 feet above him. It soon became clear to him that the USS *America* would keep its course into the Gulf of Oman until his absence was discovered at morning muster. The LCpl took a moment to consider his situation. Since land was at least 100 miles away, swimming was suicide; he would have to stay afloat until a search party found him.

Using the techniques he had learned during combat water survival training, LCpl Mayo made a flotation device out of his coveralls and tried to relax. Meanwhile, business continued as usual aboard the USS *America.* Since LCpl Mayo was on special assignment with the hazardous materials division, his absence went unnoticed until a petty officer asked several Sailors if they had seen their shipmate recently. By the time a roll call had been completed, LCpl Mayo had been adrift at sea for over 24 hours. Although three, fixed-wing Viking aircraft were deployed to search for LCpl Mayo, most people aboard ship feared the worst.

After 34 hours at sea, LCpl Mayo was discovered by fishermen on a Pakistani fishing boat. LCpl Mayo's survival is a testament not only to his incredible physical courage, but also to the soundness of the lifesaving training and techniques he received during combat water survival training.

Chapter 1
Survival at Sea

As a Marine, you face a variety of potential water emergencies whenever you cross expanses of water: ships, watercraft, and amphibious assault vehicles (AAVs) can sink; aircraft can crash into the sea; or you can accidentally fall into the water. However, there are some basic precautionary measures you can take to protect your safety and reduce your chances of becoming a water casualty. Determine the following information as soon as you board any type of vessel. Your knowing the following information may save your life or the life of your fellow shipmates.

- How many life preservers and lifeboats/rafts are on board?
- Where are the life preservers and lifeboats/rafts located?
- What type of unit survival equipment is on board?
- Are individual survival kits issued to each person on board?
- How much food, water, and medicine do the survival kits contain? When was the last time the contents were inspected for proper quantities and shelf life expiration?
- Is there sufficient survival equipment available for the number of personnel?
- How many other personnel are there on board, and where are they located?
- What are the egress procedures for the ship, boat, watercraft, AAV, or aircraft?

Abandoning Ship

When you embark on a Navy ship, you will receive abandoning ship instructions from Navy personnel. If given the order to abandon ship, report to your designated assembly area and put on a life preserver. DO NOT inflate the life preserver until you are clear of the ship. Torn life preservers will not inflate and inflated life preservers can block you, and those behind you, from exiting the ship. A flotation device that has been inflated may also burst if you jump from a significant height. See pages 1-15 through 1-35 for staying afloat with and without a life preserver.

DO NOT remove your clothing, boots, or shoes before abandoning ship. Your trousers and blouse may be the only flotation devices available if your life preserver is faulty or becomes damaged, and your clothes can provide some insulation from the cold water. However, remove your soft cover and place it in a cargo pocket for later use. The soft cover is both lightweight and a good protection against sunburn caused by the sun's rays reflecting off the water.

Jettisoning Equipment

Equipment should be kept properly packed and waterproofed in case you have to abandon ship. If entering the water from a height greater than 30 feet, wearing your equipment (e.g., pack, helmet, gas mask) could cause injury. Upon impact with the water, the helmet will "cup" air inside of it. The chin strap may also create a "hanging effect" as you submerge from the force of the fall. Therefore, you should remove your helmet and gas mask before abandoning ship.

If you are unable to maintain buoyancy due to the amount of equipment secured to your pack and body, then jettisoning some of your equipment may become necessary. Equipment that you should always retain includes canteens of freshwater, first aid kit, soft cover, and survival kit. The survival kit should include first aid items, water purification tablets or drops, fire starting equipment, signalling items (e.g., flashlight, strobe light, chemlights), food procurement items, and shelter items. Other items in a survival kit include sunburn lotion and lip balm, knife, goggles/sunglasses, plastic bag, matches and lighter, mirror. See Marine Corps Reference Publication (MCRP) 3-02F, *Survival*, for a detailed list of survival items and applications.

Abandoning Ship Technique

When abandoning ship, safety considerations must be observed. Use the following technique when abandoning ship without your combat gear:

Place your hands on their opposite shoulders, forming a criss-cross pattern.

Step to the edge of the ship's deck and check the water below for debris or survivors. If the water is clear, look straight ahead and prepare to jump. If the water is not clear, move to another location.

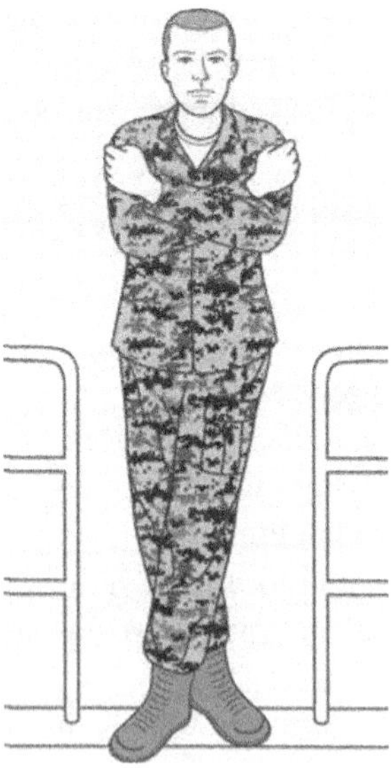

NOTE: DO NOT hold your nose as you abandon ship. If you do hold your nose, the force of impact into the water could jar your arm and hand and cause you to break your nose.

Step off the side of the ship with a smooth, 30-inch stride. DO NOT DIVE OFF THE SHIP. DO NOT LOOK DOWN AT THE WATER. LOOK STRAIGHT AHEAD. Looking down at the water can render you unconscious or cause injuries upon impact.

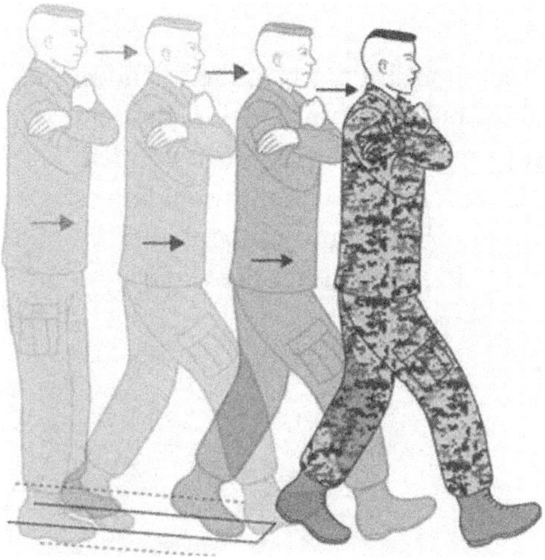

Bring your trailing leg forward during the fall. Cross your trailing leg behind your leading leg.

Keep your head parallel to the water's surface until hitting the water.

You should remain in the abandon ship position until your descent into the water has almost stopped. However, the weight imbalances in your body may cause you to be in a "J" shape under the water

.

Once your downward motion has ceased, your feet may be parallel with the ocean bottom or you may be nearly inverted with your feet over your head. To counteract potential disorientation, you should pause briefly and allow the natural buoyancy of your torso to bring your body to a nearly upright position.

Floating debris can cause hazards. Therefore, you should swim upward, extending one arm (hand is shaped as a fist) upward to feel for obstructions. If you encounter debris, try to push it away or surface in a different location.

Swim away from the ship. DO NOT LOOK BACK AT THE SHIP. Looking back could slow your movement away from the area. Remember, your objective is to leave the area as quickly as possible because—

- Equipment and debris may be falling from or spilling out of the ship.
- Additional casualties can occur if individuals abandoning the ship fall on top of swimmers already in the water.
- Swimmers close to the sinking ship may get pulled underneath the water by the suctioning effect of the ship as it goes under.

Modified Abandoning Ship Technique

When abandoning ship while wearing full combat gear (weapon, helmet, and a properly waterproofed pack), safety considerations must be observed. The modified abandoning ship technique is used while wearing full combat gear and exiting from a height less than 30 feet.

WARNING

When exiting from a height greater than 30 feet, remove your helmet and pack. Fasten the helmet to the pack or place it inside the pack before jettisoning. If jettisoning gear from the ship or aircraft, check the water below for survivors before throwing the gear forward of the intended jump area. Once in the water, retrieve your gear and swim out of the area.

Place your weapon over one shoulder, muzzle down, with the weapon parallel to your side.

Place your arm and hand along the weapon and hold it to your side.

Take your free hand and place it on top of your helmet to prevent neck/spinal injury from the force of the water pulling upward on the helmet as your body enters the water.

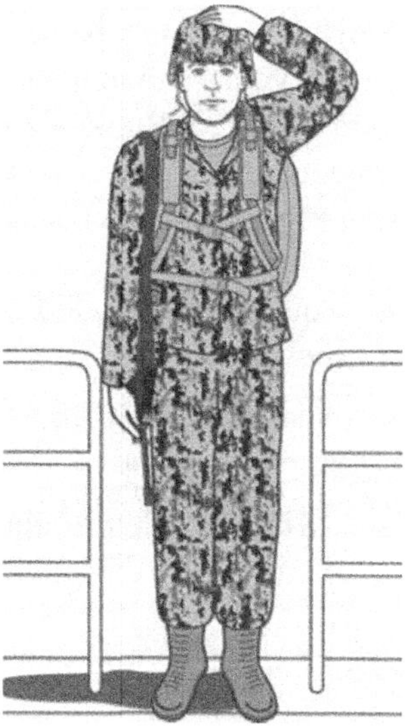

NOTE: An alternate method is to place your weapon over one shoulder, muzzle down, with the weapon parallel to your side. Reach across your body and grasp the sling of your weapon and hold it to your body. Take your free hand and place it on top of your helmet.

Step to the edge of the platform and check the water below for debris and survivors. If the water is clear, look straight ahead and prepare to jump. If the water is not clear, move to another location.

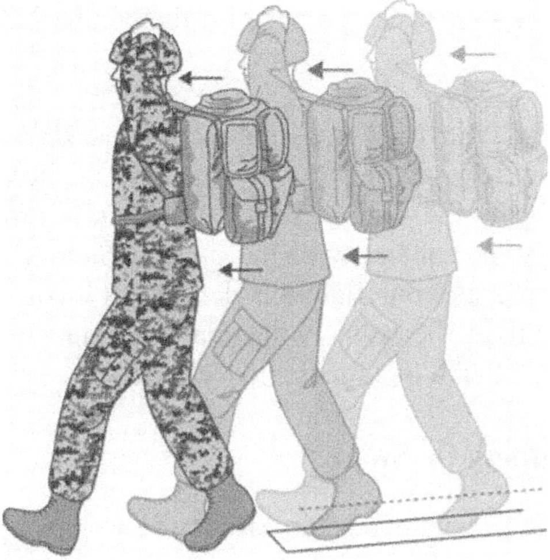

Step off the side of the platform with a smooth, 30-inch stride. DO NOT DIVE OFF THE SHIP, DO NOT LOOK DOWN AT THE WATER, LOOK STRAIGHT AHEAD. Looking down at the water can render you unconscious or cause injury upon impact.

Bring your trailing leg forward during the fall. Cross your trailing leg behind your leading leg.

Keep your head parallel to the water's surface until hitting the water.

You should remain in the modified abandoning ship position until your descent into the water has almost stopped. The buoyancy of a properly waterproofed pack will immediately pull you to the surface. Once you break the water's surface, unsling your weapon and loop the sling over your head. You may have to seesaw the sling so as to ensure the sling passes between your helmet and pack. Once over your head, loop the sling around your neck with the weapon aligned in the center of your body, muzzle down. Lean back on your pack and perform the combat travel stroke to exit the area.

> *NOTE:* You may also remove the pack (unsnap the quick-release strap on one side of the pack, and pass the strap to your other hand in order to maintain contact with the pack, which is functioning as your flotation device). Remain horizontal in the water with your pack under your chest. Perform a breast stroke kick.

Surface Burning Oil Swim

After you have abandoned ship, rise to the surface using the techniques shown on the page 1-6. However, you must remember that fuel from sinking ships or downed aircraft will float on the surface of the water. Therefore, you must move clear of the floating fuel by swimming away from the ship or aircraft as soon as possible. Either swim upwind (into the wind) of the ship/aircraft or swim against the current. Either method allows you to move away from the fuel and the wind/current will push the fuel past you. To properly execute a surface burning oil swim—

Extend your arms overhead as far as possible.

Wave your arms back and forth vigorously to splash a hole while moving upward.

Splash as long as possible to push burning fuel away from the surfacing area.

Use your arms and hands to sweep away fuel and debris.

Kick your legs in a constant breast stroke kick.

Extend your arms (palms outward) forward on the surface, arms shoulder-width apart.

Pull your hands in and back toward the chest.

Stop your hands in front of your face and rotate them so that your palms face forward (roughly halfway out of the water).

Sweep your arms forward to a full extension at the shoulder width. This splashes debris, oil, or burning liquids aside. To reduce the chance of fatigue, use two short splashes to the front to extend the path.

Repeat the preceding step as necessary while swimming clear of the area.

Surviving With a Pack

If packed properly, your pack will float, and it is your key piece of equipment for staying afloat and overcoming water obstacles. If the pack's contents are properly waterproofed, it can support you (with a combat load) in the water. Buoyed up by a waterproofed pack, you eventually emerge from the water with all your equipment (e.g., boots, helmet, flak jacket, weapon, survival items).

Your pack floats based on a scientific principle known as Archimedes' principle. This principle states that an object submerged in a liquid is buoyed up by a force equal to the weight of the liquid displaced (pushed aside) by the object. If the weight of the displaced liquid is greater than the weight of the object, the object floats. If the weight of the displaced liquid is less than the weight of the object, the object sinks. For example, a machine gun sinks in the water, but it still weighs less in the water than it does on land. Even though a machine gun sinks, it is still buoyed up by a force equal to the weight of the water it displaces. For this reason, you should not try to hold yourself or your equipment any higher out of the water than they would naturally float; doing so wastes both energy and body heat.

Preparing Equipment

Before packing your pack, you must prepare your gear/equipment. Tape or pad all sharp edges and equipment corners. Ideally, your gear/equipment is placed in plastic bags and the plastic bag is then placed inside the standard-issued, rubberized, waterproof bag. Waterproof bags are not completely water tight; henceforth, the added protection of first wrapping the gear/equipment in a plastic bag, then placing the plastic bag inside the waterproof bag. Large plastic bags (e.g., trash bags) work well for bulky equipment (e.g., sleeping bags, field jackets, shelter halves, gas masks). Small plastic bags work best for small items (e.g., shaving gear).

> *Note:* If a gas mask must be carried outside the pack, cover it with a waterproof bag.

You will need the following items, which are available through the supply system, to prepare your equipment for packing:

- Waterproof bags:
  Pistol bag, plastic size #1 (8" x 18").
  Rifle bag, plastic size #2 (10" x 56").
  Machine gun bag, plastic size #3 (15" x 56").
  Multipurpose bag, cover, plastic size #4 (20" x 84").

- Panama work vest (life preserver).
- Small plastic bag, self sealing (6" x 6" and 12" x 12").
- Riggers tape, olive drab (2 1/2").
- Willie peter bag, waterproof clothing.

Tying Waterproof/Plastic Bags

Try to remove the excess air from the waterproof/plastic bag before securing the bag's opening. If filled with air, the bag can burst if pressed from the outside. Perform the following steps to tie a waterproof/plastic bag:

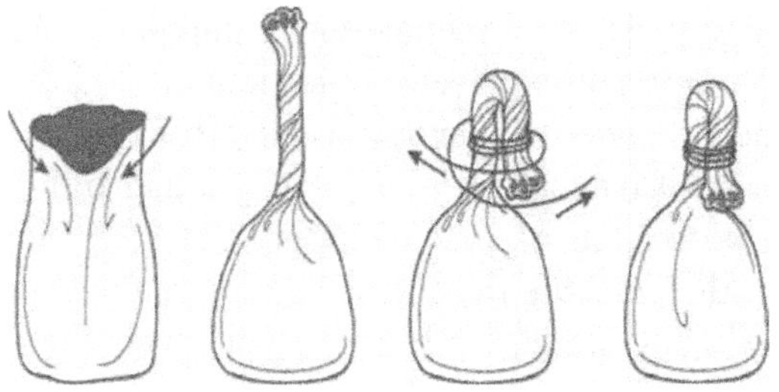

Packing the Pack

Place filled and tied bags inside the pack. Carefully handle sharp items (e.g., tent pegs, poles) to prevent puncturing the bag. Place items in the pack in order of expected use. Close the pack and its compartments. Attach sleeping mats or bags as high as possible on the outside of the pack.

Swimming With the Pack

A useful technique for propelling yourself forward while wearing a pack is the combat travel stroke. This technique allows you to float nearly horizontal and to propel yourself forward with bicycle-style kicks and breast stroke-style arm sweeps. To execute the combat travel stroke—

Body position
Keep the upper part of your body prone to the water, let your legs dangle horizontally, and keep your face up.

Arm action
Extend your hands out in front of your waist. Sweep your arms slightly downward and back 90 degrees to propel your body through the water. Move your hands back to the front of your waist. Repeat.

Leg action
Continuously move your legs in a bicycle-like movement, bringing your knees up high and step out.

Breathing
Keep your face out of the water during the stroke, and breathe freely.

Staying Afloat With a Life Preserver

The best form of flotation is to find any kind of floating object that will keep you and your equipment out of the water or minimize your exposure to the water. Life preservers are the best method, as they allow you to wear your clothes for heat retention and sunburn prevention. Marines use two basic classes of life preservers: inherently buoyant life preservers and inflatable life preservers.

Inherently Buoyant Life Preservers

Inherently buoyant life preservers are either vest-type (worn like a jacket) or yoke-type (worn around the neck). The preserver's outer envelope is either a cotton or water resistant material that encloses a removable fibrous glass or plastic foam filling.

In the Marine Corps, the most common type of inherently buoyant life preserver is the vest-type with collar, known as the kapock preserver. The kapock consists of collar straps, upper front chest straps, leg straps, and waist drawstrings that secure the preserver to you. The leg straps, which are fitted on both sides of the life preserver, ensure that the preserver remains around your chest while you are in the water. A chest strap is attached to the life preserver to facilitate lifting you out of the water. The strap can also be attached to other survivors or to lifeboats to reduce the fatigue that results from holding onto a floating/secured object by hand.

Inflatable Life Preservers

Marine Corps aircraft and AAVs have inflatable life preservers on board. The inflatable life preserver issued to Marines is known as the LPP (life preserver personal). The LPP is capable of both oral inflation and CO_2 cartridge inflation. The LPP consists of buoyancy chambers, CO_2 inflator, and an oral inflation tube. The buoyancy chamber is deck gray in color and is made from a neoprene-coated nylon fabric.

Inflatable life preservers must be stored in a cool, dry place. Heat, moisture, and light cause deterioration of the life preserver material. Do not stow CO_2 cylinders near steam lines or radiators. Heat can increase the pressure inside the cylinders causing them to explode. Avoid sharp edges in stowage. Sharp edges increase wear and tear on the life preservers and may also puncture inflatable buoyancy chambers.

> **CAUTION:** Do not inflate the life preserver until you are clear of the aircraft, ship, or vehicle. Torn life preservers will not inflate and inflated life preservers can block you, and those behind you, from exiting the aircraft, ship, or vehicle.

To don and adjust the LPP—

Remove the LPP from its storage container.

Fasten the belt fasteners in front with the pouch in the rear.

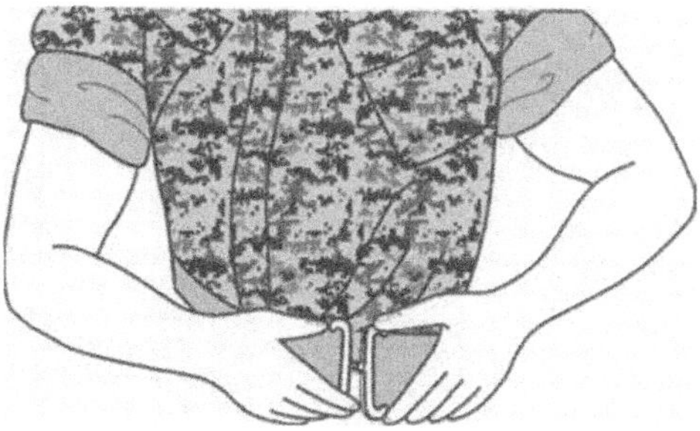

Adjust the belt to fit; secure any excess belt by mating the hook and pile tape.

Rotate the pouch to the front and re-adjust the belt if necessary.

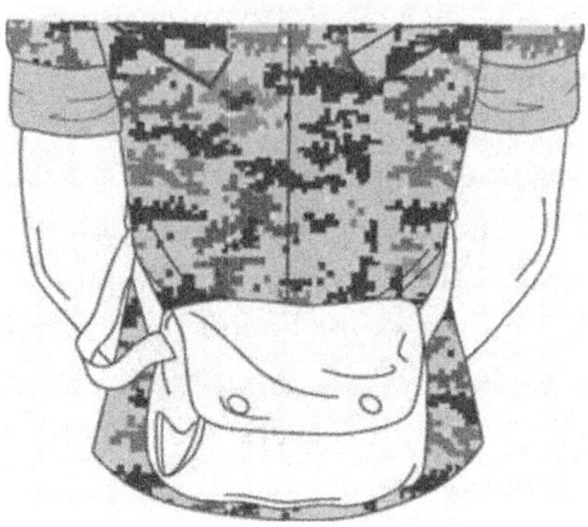

Open the snap fasteners on the pouch and unfold the life preserver.

Place the deflated life preserver over your head.

Place the storage container into the pouch after donning the life preserver.

Lift the lower end of the life preserver out of the pouch.

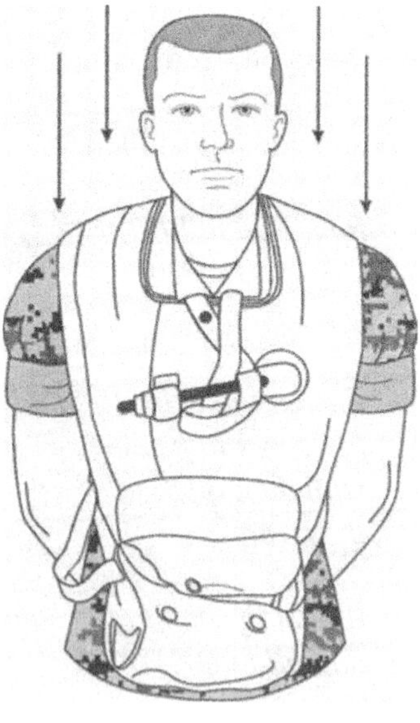

Inflate the life preserver by pulling on the lanyard attached to the CO_2 inflation valve or by blowing on the end of the oral inflation valve.

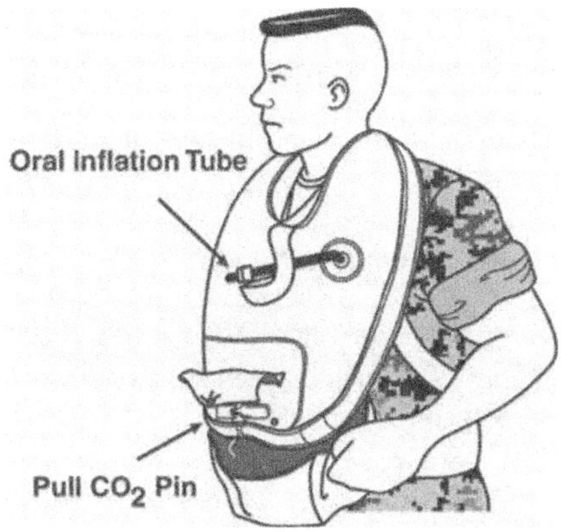

Staying Afloat Without a Life Preserver

You may be in open water without any floating objects or a life preserver to help you survive. If so, uniform blouses and trousers can be made into expedient flotation devices.

Floating With an Inflated Blouse

It is possible to float by a bubble of air trapped in the shoulders of your blouse. The air rises to the back and shoulders of the blouse and supports you at the water's surface. An inflated blouse is also a temporary flotation device used by weaker swimmers while trying to remove their trousers. There is a primary and an alternate way to create a bubble of trapped air in a blouse—

Primary Method
Turn the collar inside the blouse to help create a seal.

Unbutton top button and pull collar around mouth and nose.

Take a deep breath and bend forward slightly at the waist. Exhale one-half to three-quarters of a breath into the blouse.

Grasp and twist the collar with one hand to create a seal, this prevents air from escaping out from the collar.

Use your free hand and feet to stroke and kick to the surface.

Gather and hold the blouse tightly at the collar and stomach level to prevent the blouse from losing air if it floats up too high.

Splash water on the blouse periodically to prevent the material from drying, dry material allows air to escape.

Repeat inflation as required.

Alternate Method
Turn the collar inside the blouse to help create a seal.

Unbutton the second button from the top.

Take a deep breath and bend forward slightly at the waist.

Place your mouth and nose inside the hole created by the open button and exhale one-half to three-quarters of a breath into the blouse.

Grasp material at the unbutton portion and pull downward.

Use your free hand and feet to stroke and kick to the surface.

Splash water on the blouse periodically to prevent the material from drying, dry material allows air to escape.

Repeat inflation as required.

Floating With Inflated Trousers

In warm water, trousers can be used as a primary expedient flotation device. However, in cold water, submerging your head to remove and inflate your trousers results in heat and energy losses that negate the benefit of using the trousers as a flotation device. Once your trousers are inflated, you float motionlessly as if wearing a life preserver. If needed, assume the heat escape lessening posture (known as the HELP position, see page 1-37) to slow heat loss. As trousers dry, air leaks out of the legs. To slow this process, occasionally splash water on the fabric. Reinflate trousers as needed.

Sling Method

The sling method works if you are a strong swimmer or naturally very buoyant. Take the following steps to inflate trousers using the sling method:

Take a deep breath, bend over, and remove your boots.

NOTE: Retain your boots. Tie the boot laces together and suspend the boots from your blouse or hang them around your neck so that they rest on your chest.

Remove your trousers. Button or zip the trouser's fly closed. This allows you to control airflow.

Tie the bottoms of the trouser legs in a square knot. (The appendix illustrates various knots.)

Ensure that the front (fly) of the trousers faces you.

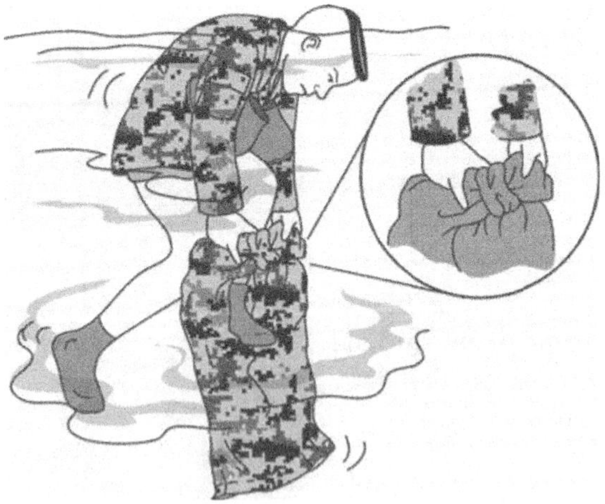

Hold the trousers above the water's surface and behind your head. Grasp both sides of the waistband and open with both hands.

Kick strongly to stay on top of the water while slinging the trousers overhead in order to trap air into them.

Once the waistband is submerged in the water, air is trapped in the legs.

Hold and seal the waistband underwater.

Slip the inflated legs over your head. Hold the waistband in toward your chest, the fly facing your body. To prevent air from escaping from the trousers, seal the waistband by either folding it or twisting it.

Lie back and relax, resting the back of your neck against the knot.

Splash water on the trousers periodically to prevent the material from drying. Dry material allows air to escape.

To replenish air in the trousers, you will use a technique known as the scooping method. With one hand on the open waistband, extend the trousers in front of you just below the surface of the water and scoop air bubbles with your free hand into the open waistband until the trousers have sufficient air. Repeat as necessary.

Splash Method

The splash method is an alternative to the sling method. As with the sling method, you must kick strongly to remain at the surface. To inflate trousers using the splash method, perform the following:

Take a deep breath, bend over, and remove your boots.

NOTE: Retain your boots. Tie the boot laces together and suspend the boots from your blouse or hang them around your neck so that they rest on your chest.

Remove your trousers. Button or zip the trouser fly closed. This allows you to control airflow.

Tie the bottoms of the trouser legs in a square knot. (The appendix illustrates various knots.)

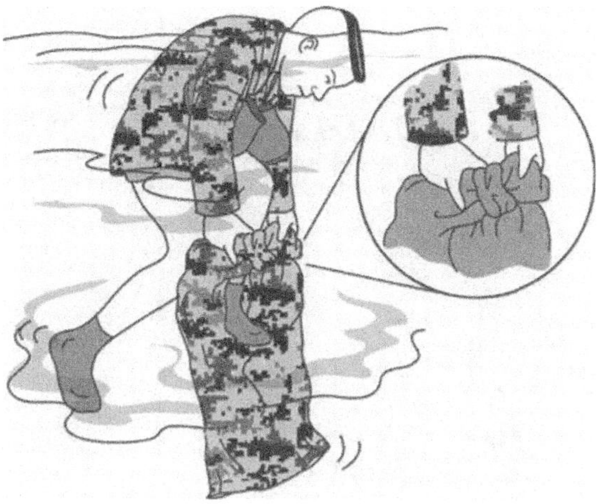

Ensure that the front (fly) of the trousers faces you.

Hold the trousers at the water's surface out in front of you by the waistband with the fly up.

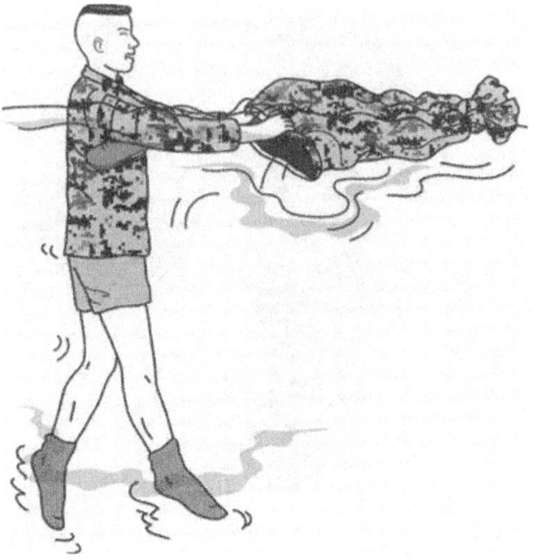

Grasp the waistband at the surface with one hand. Insert your free hand into the waistband, palm down.

Flutter your hand rapidly to create bubbles. This sends a mixture of water and air bubbles into the trousers. The water passes through the fabric. The air remains trapped in the legs.

Hold and seal the waistband underwater.

Slip the inflated legs over your head. Hold the waistband in toward your chest, the fly facing your body. To prevent air from escaping from the trousers, seal the waistband by either folding it or twisting it.

Lie back and relax, resting the back of your neck against the knot.

Splash water on the trousers periodically to prevent the material from drying. Dry material allows air to escape.

To replenish air in the trousers, you will use a technique known as the scooping method. With one hand on the open waistband, extend the trousers in front of you just below the surface of the water and scoop air bubbles with your free hand into the open waistband until the trousers have sufficient air. Repeat as necessary.

Blow Method

The blow method is an alternative to the sling method. Use the blow method if you are a weak swimmer. Take the following steps to inflate trousers using the blow method:

Take a deep breath, bend over, and remove your boots.

NOTE: Retain your boots. Tie the boot laces together and suspend the boots from your blouse or hang them around your neck so that they rest on your chest.

Remove your trousers. Button or zip the trouser fly closed. This allows you to control airflow.

Tie the bottoms of the trouser legs in a square knot. (The appendix illustrates various knots.)

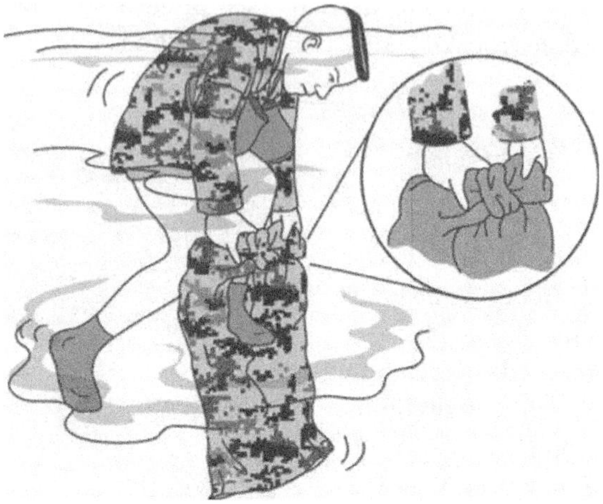

Ensure that the front (fly) of the trousers faces you.

Hold the trousers at the water's surface. Grasp both sides of the waistband and open with both hands.

Take a deep breath.

Drop 2 feet below the water's surface, pulling the waistband underwater.

Hold the waistband open with both hands and blow air into the trousers.

To fill the trousers with air, surface while keeping the waistband underwater, breathe in again, drop below the water's surface, and blow air into the trousers. Repeat these steps until the trousers are filled sufficiently. Once trousers are filled—

Hold the waistband underwater. Twist and pinch it off.

Slip the inflated legs over your head. Hold the waistband in toward your chest, the fly facing your body. To prevent air from escaping from the trousers, seal the waistband by either folding it or twisting it.

Lie back and relax, resting the back of your neck against the knot.

Splash water on the trousers periodically, to prevent the material from drying. Dry material allows air to escape.

To replenish air in the trousers, you will use a technique referred to as the scooping method. With one hand on the open waistband, extend the trousers in front of you just below the surface of the water and scoop air bubbles with your free hand into the open waistband until the trousers have sufficient air. Repeat as necessary.

Avoiding Heat Loss in Cold Water

The rate of heat exchange in the water is about 25 times greater than it is in air of the same temperature. When you are immersed in cold water, hypothermia occurs rapidly due to the decreased insulating quality of wet clothing and as a result of water displacing the layer of still air that normally surrounds the body. You also lose about 50 percent of your body heat through your head; therefore, keep your head out of the water. Other areas of high heat loss are the neck, the armpits/sides, and the groin.

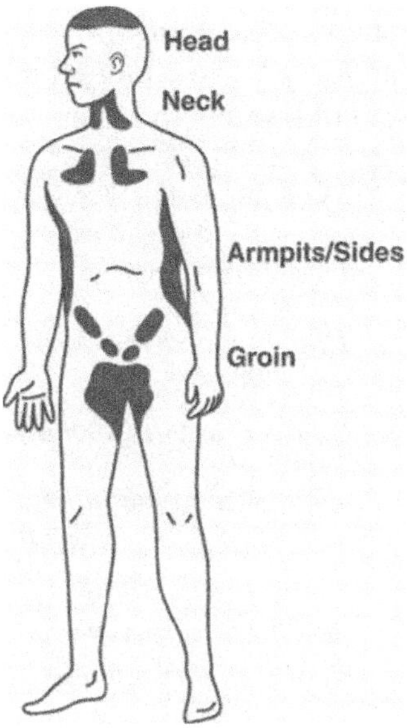

In cold water, DO NOT SWIM TO STAY WARM. Swimming, even with a slow and steady stroke, produces a lot of heat that is lost in the water. The heat loss can produce hypothermia that slows body functions and can result in serious injury or death. Remaining motionless conserves body heat three times longer than swimming. SWIM only if you have a flotation device and the shoreline is visible.

Individual Protection From the Cold

If you are equipped with a life preserver, assume the heat escape lessening posture (known as the HELP position) to slow heat loss and to protect major blood vessels near the body's surface. These areas lack insulating fat and are vulnerable to the chilling effects of cold water. To assume the HELP position—

Tuck your chin down tightly to cover your throat.

Draw your legs up in a fetal position to protect the groin.

Place your arms across your chest, tuck your hands into your armpits.

Wear some type of head covering (e.g., stowed cover, towel, handkerchief) to lessen heat loss through the scalp if head covering is available.

Group Protection From the Cold

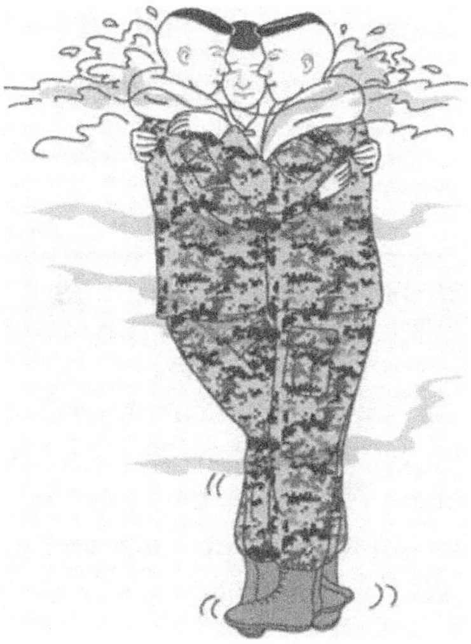

If three or more Marines are in the water and are equipped with life preservers, they should wedge tightly together and lock arms to form a circle known as a huddle position. This position protects vulnerable areas from heat loss.

A casualty who is suffering from the effects of the cold can be placed within the huddle to be surrounded by warmer water. If in the water for a prolonged period, it is recommended that everyone be rotated inside the huddle to maintain or re-warm each person's internal core temperature. If there are more than five Marines, they should make clusters of huddle positions.

Contact with other swimmers provides survival advantages—

Creates a larger target for search and rescue aircraft.

Provides additional warmth in cold water.

Improves morale.

Re-establishes the chain of command.

Reduces shock and panic.

Provides opportunities to administer first aid.

Supports exhausted Marines.

Drownproofing Methods

An object that floats on the surface has positive buoyancy. An object that floats a few feet beneath the surface has neutral buoyancy. An object that sinks has negative buoyancy. Most people have positive buoyancy and will float at the water's surface. To test your buoyancy—

Stand in water that is at your chest level or deeper.

Take a full breath.

Bend forward slowly.

Relax and wait.

If you have positive buoyancy, you will slowly rise to the surface. If you have neutral buoyancy, you will float a few feet beneath the surface. If you have negative buoyancy, you will sink. Regardless of whether you naturally float or sink, you can remain at the surface for extended periods without a life preserver if you exercise the appropriate drownproofing methods or survival strokes, which are based on your buoyancy.

Drownproofing methods consist of the T-method and the sweep. The breast stroke, side stroke, and elementary backstroke are the most common survival strokes. With any drownproofing method or survival stroke, remember the acronym **SAFE**:

SLOW EASY MOVEMENTS
Move slowly to conserve energy and minimize heat loss.

APPLY NATURAL BUOYANCY
Use natural buoyancy to support the body.

FULL LUNG INFLATION
Fill the lungs with each breath. Do not hold air in the cheeks.

EXTREME RELAXATION
Tight muscles are denser than relaxed ones and do not float as well.

Crawl Stroke

The crawl stroke, sometimes called the front crawl or free style, is the fastest stroke. To execute the crawl stroke—

Body position
Lie horizontal, on your stomach, in the water.

Look forward and downward at a 45 degree angle with the waterline between your eyebrows and hairline. Your head position is important as it assists in cutting a path through the

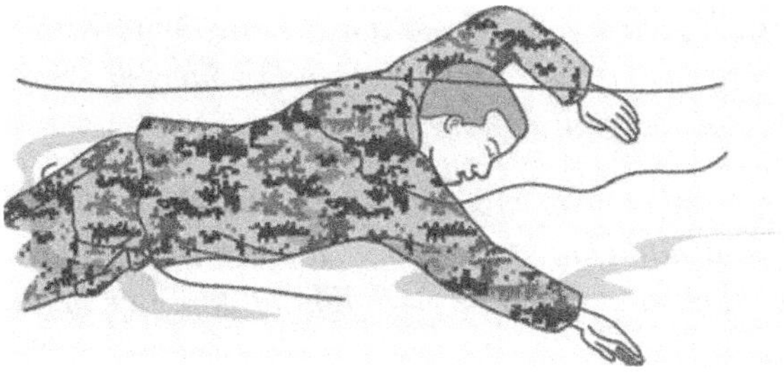

water. If your head is too high in the water, your lower body sinks significantly, making your stroke less efficient. If your head is too low in the water, water washes over your shoulders and neck, causing unnecessary drag.

Arm action

Arm action occurs in three phases: catch, propulsion, and recovery.

Fully extend one arm forward of your body, this positions your hand to catch the water in preparation for the propulsion phase.

To catch the water, bend your wrist (with your palm pointing outboard) and make an "S" shape (or inverted "S" shape) with your hand, ensuring that your hand does not cross the center of your body. Your left hand makes an "S" shape and your right hand makes an inverted "S" shape with your hand finishing at shoulder level.

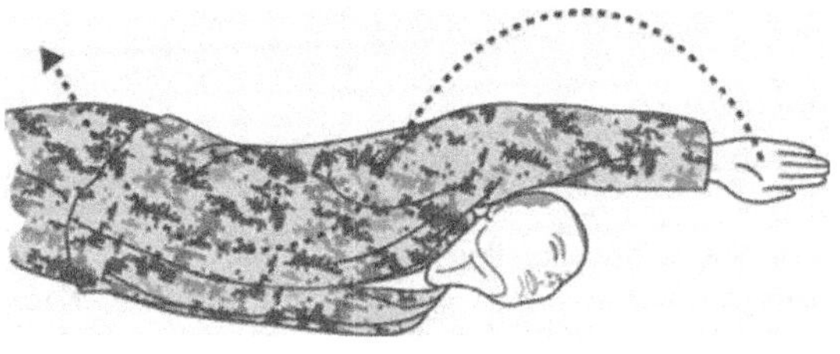

Push with your hands in a rearward fashion toward your feet until your arm is fully extended along your side, keeping your hands close to your body.

To begin the recovery phase, bend your arm at the elbow and raise your hand out of the water. Your hand breaks the surface of the water and maintains a height of 2 to 3 inches above the water's surface.

With your hand and arm moving in a forward manner, bring your hand past your head until your arm is about three-quarters of the way extended. At this point, turn your hand with your palm outboard so as to allow your thumb and fore-finger to enter the water first.

Once your hand enters the water, continue to push your arm forward until it is fully extended.

You are now prepared to catch the water again. These steps are performed in an alternating pattern: when one arm is catching and propelling, the other arm is recovering.

Leg action
Use a flutter kick to create the leg action for the crawl stroke. This kick is used for both propulsion and keeping the lower body horizontal with the water's surface. The flutter kick is an alternating leg action: one leg is kicking in a downward motion while the other leg is recovering to the surface to prepare for the next kick. The size of the kick ranges from 12 to 15 inches and depends on your height.

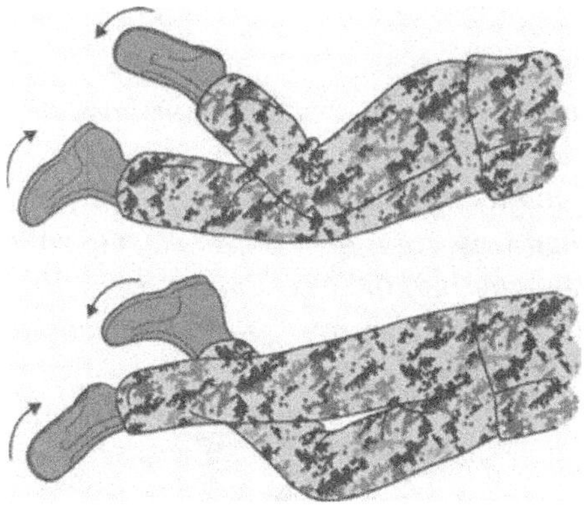

Maintain your legs in a semi-rigid manner.

Generate power for the kick from the hips.

Keep your feet loose at the ankles, they trail behind your legs and act as "flippers."

Execute the propulsion phase of this kick with a downward thrust of your leg.

Execute the recovery phase by pushing your leg back to the surface. This phase is complete when your foot reaches the surface.

Breathing
Breathe during the recovery phase of the arm action.

Roll your body and rotate your head to the side of your body where the arm recovery is occurring, this rolls the water away from your mouth. Keep your chin pushed back toward your shoulder.

Exhale while your face is still submerged and inhale when your face breaks the surface.

Breathe either bilaterally or rhythmically. To breathe rhythmically, breathe on the same side of your body every time your arm cycle occurs. Bilateral breathing requires you to breathe every one and a half arm cycles. To breathe bilaterally, breathe when your right arm is recovering, your face goes back in the water and the next time you breathe is when your left arm is starting its recovery three arm strokes later.

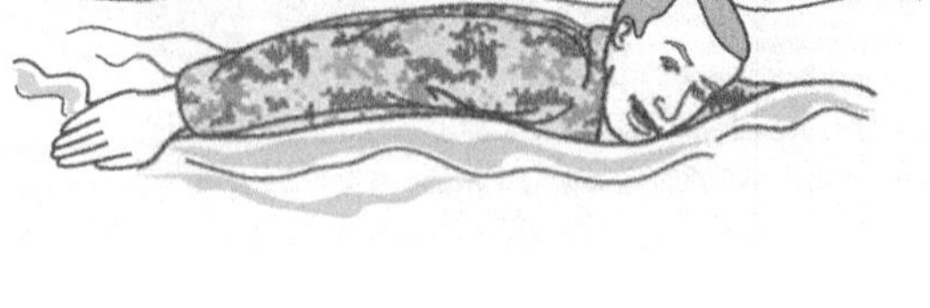

Bilateral breathing is the preferred method for breathing; it prevents the chance of hyperventilation and allows your body to maintain a lateral position in the direction you are swimming.

Coordination

This stroke uses constant arm and leg action.

T-Method

The T-method is a basic drownproofing method. This is the best survival technique if you have negative buoyancy. To execute the T-method—

With your face out of the water, take a deep breath and submerge your face in the water while holding your breath.

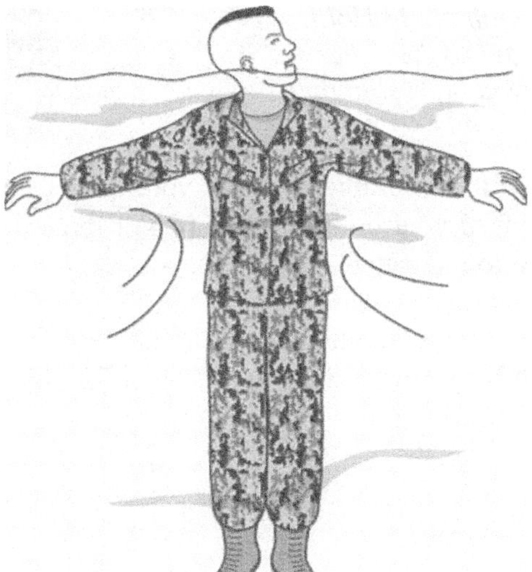

Float with your body in a horizontal position, arms extended from your side, and legs extended and joined.

Move your hands up to your armpits by tracing an imaginary line along your ribs.

Extend your arms outward (horizontal) to your sides, your body position resembles the letter "T."

Step out and forward with one leg and point your other leg to the rear, your knees should be slightly bent. Simultaneously, bring your arms down to your sides. Then bring your legs back together. You complete the steps by exhaling most of your air and preparing to surface your face to obtain another breath of air. Hold your head out of the water. Tilt your head back slightly. Breathe normally.

Once your breath is complete, move your hands up and down directly in front of your body. Do this two or three times to slow your descent into the water.

> NOTE: To avoid hyperventilating, hold your breath below the surface of the water for no more than 10 seconds.

The Sweep

The sweep works well if you have slight to excellent positive buoyancy. To execute the sweep—

Float face down in the water: bend 45 degrees at the waist, arms and legs dangling, head hanging down, relax all muscles.

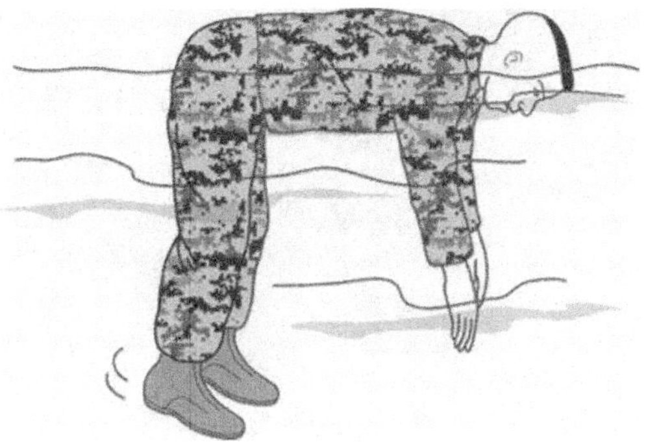

Spread your feet slowly to prepare for a single kick (one leg is forward and one leg is rearward).

Cross your arms in front of your chest, palms outboard with the back of each hand touching the opposite ear.

Exhale prior to raising your head for a breath.

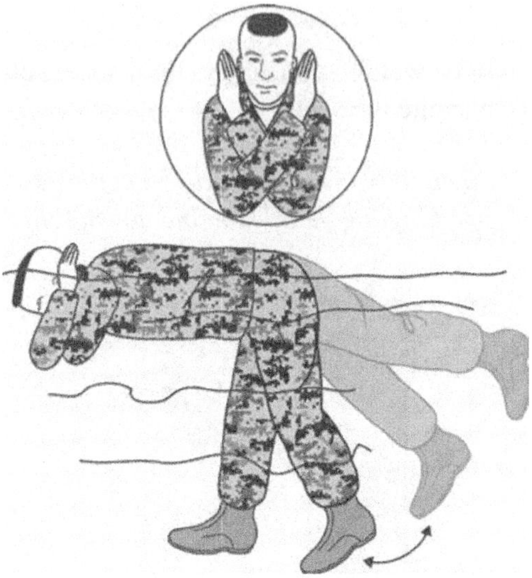

Bring your legs together and sweep your arms down and out until the arms are fully extended out to the sides. This raises your face above the water and allows you to catch a breath of air.

With your air supply replenished, return your face to the water and relax while sweeping your arms in a downward motion in front of your body to prevent/slow your descent. Do not hold your breath for more than 10 seconds. More than 10 seconds enhances your chances of shallow water black out and subsequent drowning.

Breast Stroke

Use this stroke to swim underwater, through oil or debris, and in rough seas. If you are a good swimmer and not wearing combat gear, the breast stroke is the best stroke for long-range swimming because it provides good visibility and allows you to conserve your energy and maintain a reasonable speed.

Body position
Lie prone in the water. Swim with your trunk and legs projecting back and down at an angle of 20 to 30 degrees.

Extend arms out in front (hands together [side by side]), and extend legs behind (toes pointed) to prevent drag.

Face downward, looking forward at a 45 degree angle to break the water and to prevent water from washing into the collar area causing drag.

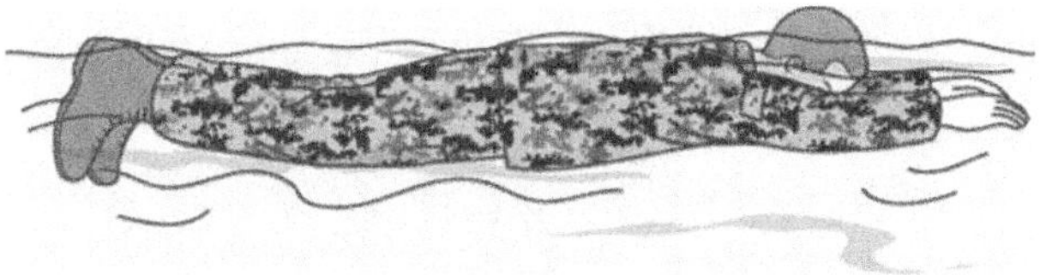

This is known as a glide.

Arm action

Turn your palms outward and bend your arms slightly.

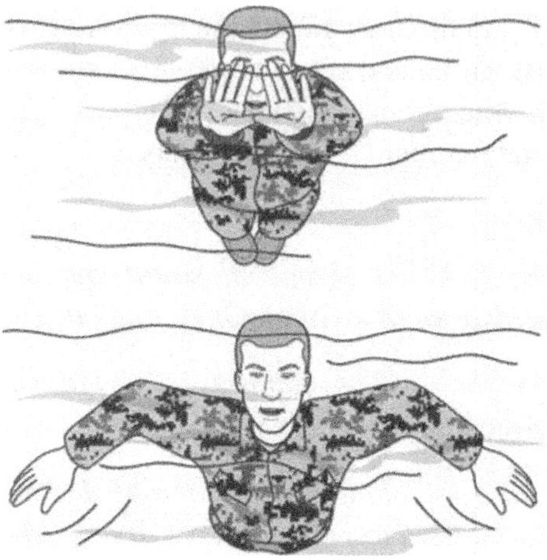

Sweep your arms sideward and slightly downward until your hands are opposite and slightly below your shoulders.

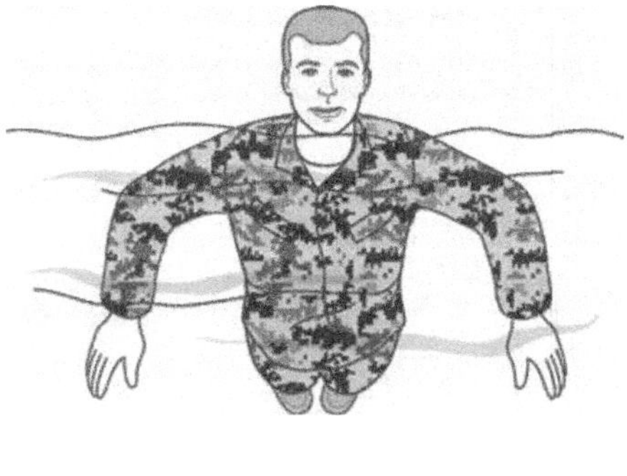

Rotate your head up, breathe once your mouth breaks the surface.

Bring your hands and arms up along your chest and thrust them forward until they are extended and ready to execute the next arm pull.

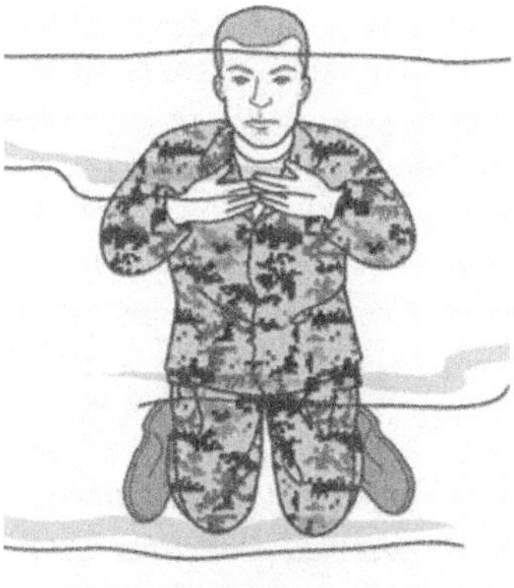

As the arms start their recovery into the glide, the head should rotate forward, resubmerging the face.

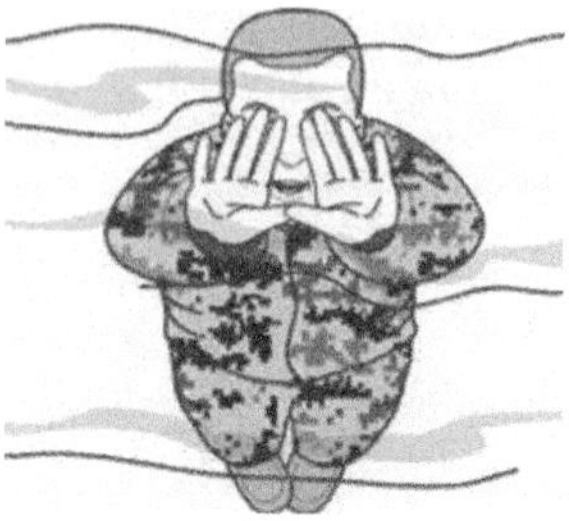

Leg action

Draw your heels toward your buttocks, establish a 45 degree bend in the knees.

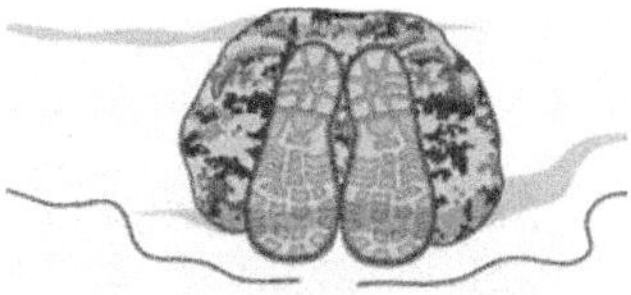

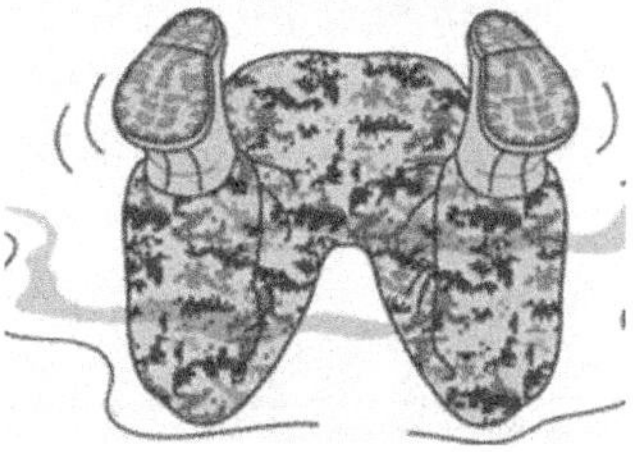

Thrust your legs outward and rearward, then squeeze them together. The whipping action of the feet aids forward propulsion.

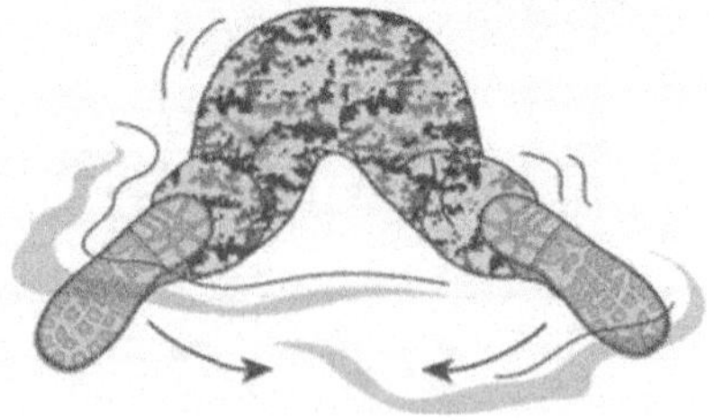

This is known as the breast stroke kick.

Breathing
Inhale during the arm pull and exhale through your mouth and nose during the finish of the breast stroke kick and glide.

Coordination
The stroke movement is in three counts:

Begin your arm pull. Near the finish of the pull, flex your knees and bring your heels toward your buttocks. The arm pull counteracts the resistance created by the knees.

As the arm pull is completed, thrust your hands forward, kick your legs outward and rearward, and squeeze them together.

Glide through the water for approximately 1 to 3 seconds or until your forward momentum decreases, then begin the next stroke.

Side Stroke

The side stroke is a survival stroke because you use both arms for buoyancy, with each arm creating a slight propulsion. The majority of your body's propulsion comes from your kick. To execute the side stroke—

Body position
Lie on your side with your lead (bottom) arm extended beyond (with a slight bend in your elbow) your head and in line with your body. Palm is down and your hand is submerged 6 to 8 inches.

Extend your trail (top) arm down the length of your body over your thigh.

Keep your legs straight and together, toes pointed rearward.

Keep your face out of the water, this allows for free breathing.

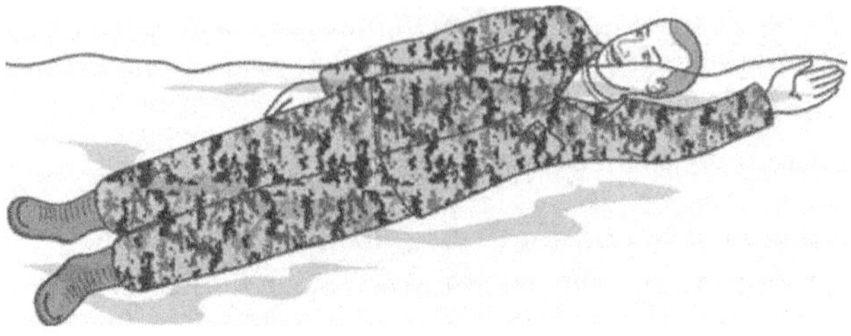

This is known as the glide.

Arm action

With your lead arm, pull your arm downward, while flexing at the elbow, until it is straight down from your shoulder.

Rotate your shoulder and pull your elbow into your side. This should put your lead hand at shoulder level. At the same time, turn your palm toward your face and thrust forward to your original, extended position.

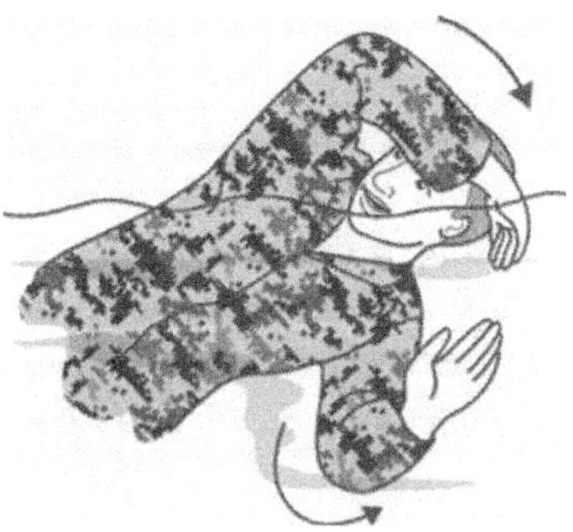

Draw your right hand upward in front of your chest to shoulder level. Rotate your palm toward your feet, then push it downward in front of your body toward your feet to catch the water.

Push your trail hand backward to its original position on top of your thigh. (Your trail hand starts forward and meets your lead hand at your chest/shoulder.)

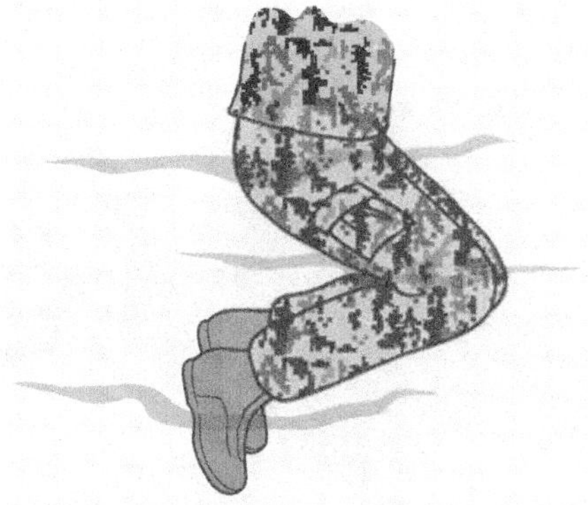

Leg action

To perform the scissor kick, the top leg always goes forward and the bottom leg always goes rearward. From the extended position, draw—or recover—your feet toward your buttocks until your legs are bent at a 45 degree angle at the knees and the hips are flexed at a 45 degree angle with the thighs.

Once the legs have completed their recovery and while maintaining a 45 degree bend in the knees, extend the legs fully into a "V" shape in order to catch the water for the propulsion phase.

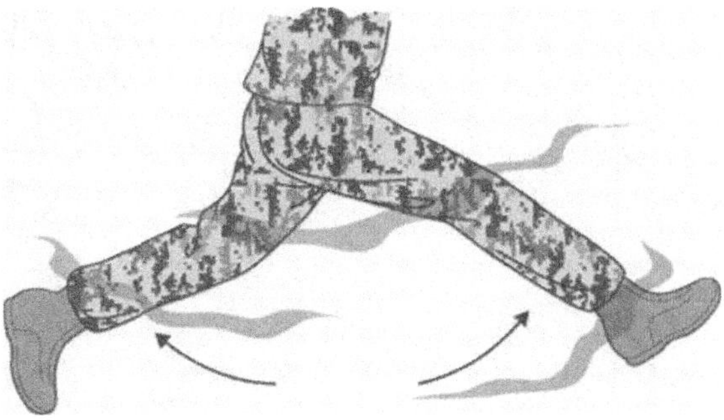

Once the legs are separated and extended forward and rearward to the "V" position, sweep the legs together until the feet are together.

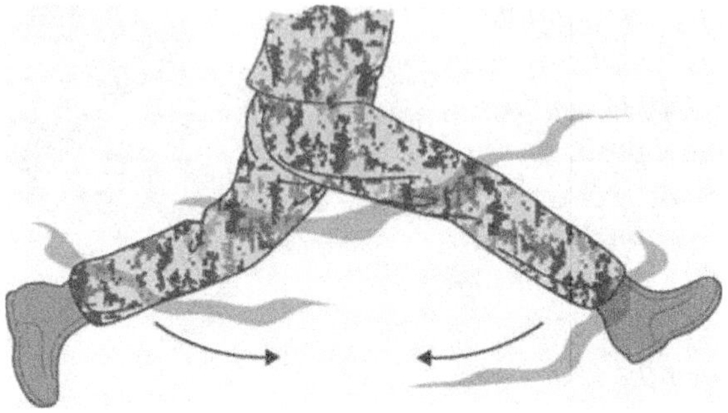

You are now in the glide position.

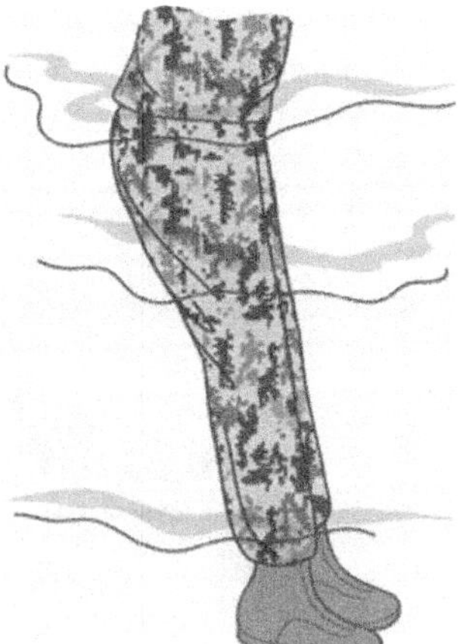

Breathing

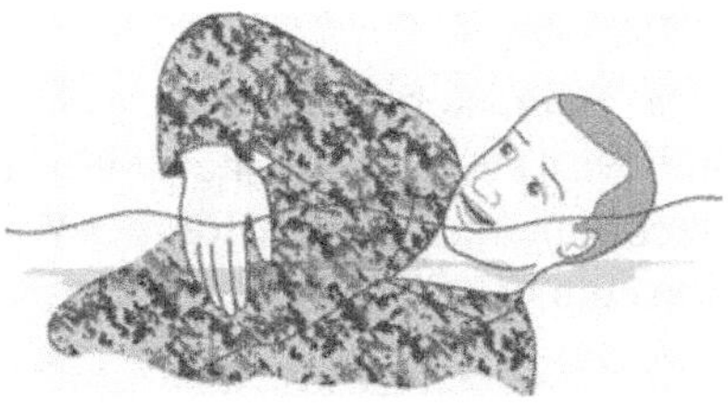

As long as your face remains clear of the water, it is a free breathing stroke. However, it is recommended that you exhale then quickly inhale when the legs are sweeping back together in the scissor kick. This is when the body reaches its highest point in the water, thus clearing the face completely from the surface of the water making it the optimum time to breathe.

Coordination

Begin the stroke with the downward pull of your lead arm. At the same time, bring your trail arm upward and draw your knees up to begin the kick. Let the thrust of the lead arm, push of your trail arm, and the kick of your legs coincide in order to finish the glide position. Glide through the water for approximately 1 to 3 seconds or until your forward momentum decreases, then begin the next stroke.

Elementary Backstroke

The elementary backstroke is also an excellent survival stroke. It relieves the muscles that you use for other strokes, and it is the recommended stroke for weak swimmers or nonswimmers. To execute the elementary backstroke—

Body position
Start on your back.

Face up, chest up, and hips up, keeping an arch in your lower back with arms pressed to your sides and your legs extended and joined to prevent drag.

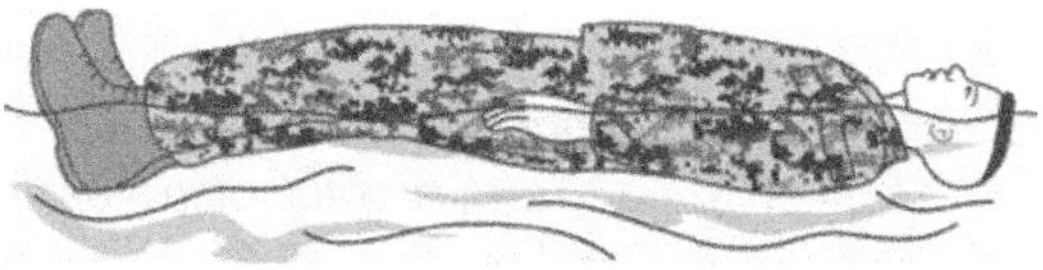

Arm action
Trace your hands up your sides to an area near your armpits then extend your arms out to the sides to form the letter "T" (palms facing feet), locking out the elbows.

> *Note:* Don't raise your arms above your head. This creates drag, changes your body position, and submerges the head.

Slap your palms to your thighs using a strong sweeping motion.

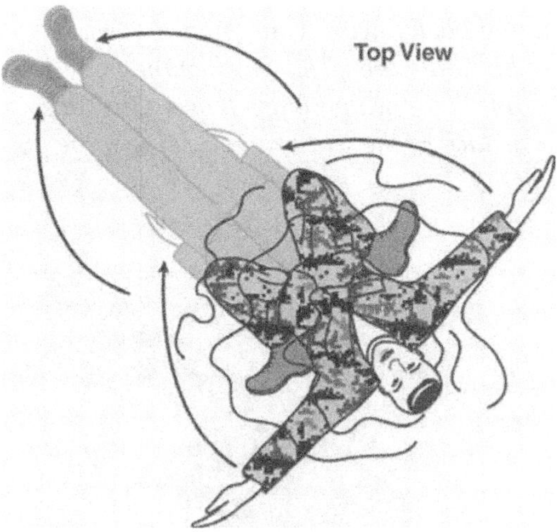

Leg action
Bend both legs at the knee (90 degree angle) slightly separating your knees and drawing your heels downward to a point under and outside your knees. The knees are spread as wide as the hips or slightly wider depending on the body type of the swimmer.

Circle around in a whipping action, ending with legs in a glide position.

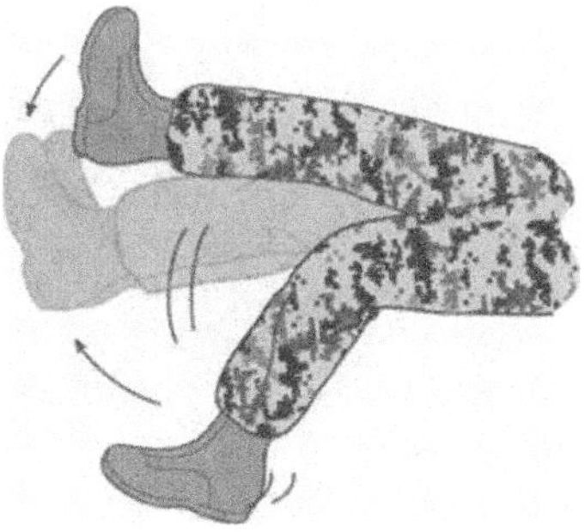

Breathing
Breathe anytime during this stroke. However, it is recommended that you exhale then quickly inhale when your arms are sweeping back toward your sides and while your legs are sweeping back together. This is when the body reaches its highest point in the water, thus clearing the face completely from the surface of the water and making it the optimum time to breathe.

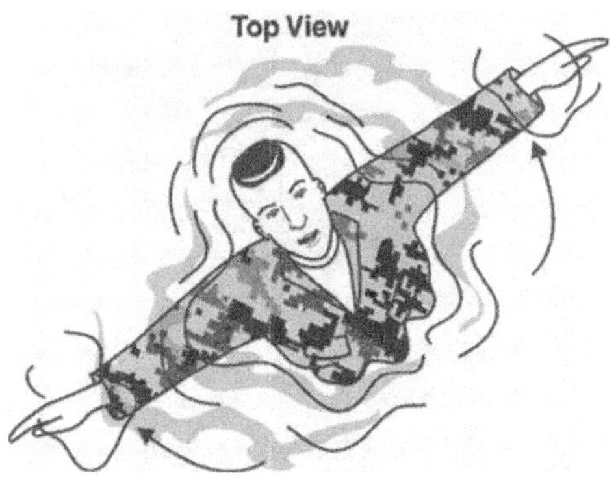

Coordination
The stroke movement occurs in three counts (recovery, catch, power).

Begin the arm pull (recovery).

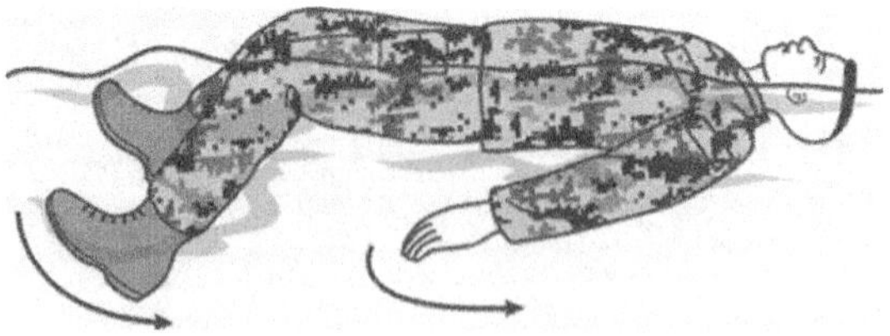

Near the finish of the pull, flex your knees to a 90 degree angle. The arm pull counteracts the resistance created by the knees.

Kick out your legs, and squeeze them together as the arm pull is completed (catch, power).

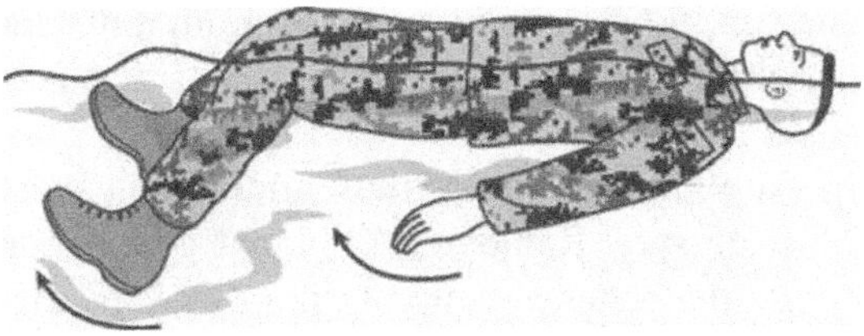

Glide through the water for 1 to 3 seconds or until your forward momentum decreases, then begin the next stroke as your momentum slows.

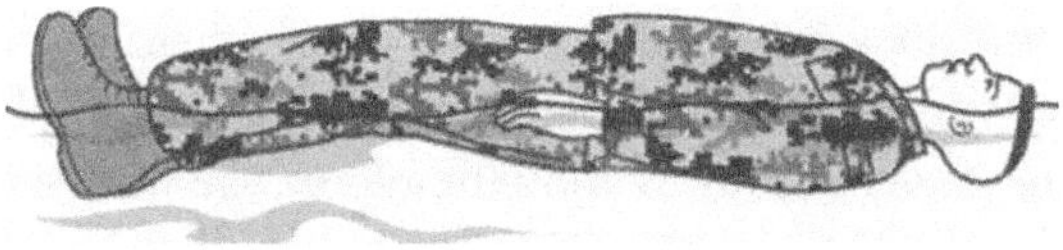

Chapter 2
Water Rescues

A drowning victim can panic and react with unexpected violence and can seize and inadvertently drown a rescuer. Therefore, if possible, a water rescue should be executed from a distance. Reaching, wading, or throwing methods are used in a distance rescue.

If a victim is too far away to use these methods, a swimming rescue may be necessary. Remove all combat gear, when possible, before entering the water. Swim within 2 to 6 yards of the victim to maintain a margin of safety, this allows you to reassess the situation and reassure the victim. If the victim is unconscious, use the wrist tow method or cross-chest carry method to pull the person to safety. If the victim is struggling, use a rear approach and then execute either a single armpit level off or a double armpit level off before towing the victim to safety.

If the victim does begin to overpower you, there are techniques that allow you to defend yourself without having to abandon the rescue. These techniques include the block, the wrist-grip escape, the front head-hold escape, and the rear head-hold escape. These techniques allow you to separate yourself from the victim, reassess the situation, and then attempt the rescue again.

> *Note:* In this chapter's illustrations, the rescuer is shown without a helmet.

Reaching Rescue Techniques

Reach

From a safe position at the water's edge, reach out to the victim. Talk constantly to calm the victim. Retain partial contact with land or some solid support structure (e.g., pier, bridge). If the victim is close but still beyond reach, extend an object (e.g., stick; pack; rifle with magazine removed, chamber empty, and muzzle pointing toward victim) that the victim can grasp. Pull the victim slowly to safety. Once the victim is close to shore, remove the victim from the water.

Note: You can also extend a foot to the victim if you can retain a secure grip on a solid support structure.

Reach From a Deck

Reach from a deck rescue can be executed by either a swimmer or a nonswimmer, and it can be used on an active or a passive victim. Be sure to reassure the victim during the rescue. To execute reach from a deck—

Lie prone on the deck with your body firmly anchored. To anchor your body, lie flat, spread legs apart, and extend one arm behind you with your palm down and on the deck.

Keep as much of your weight on the deck as possible and extend your free hand to the victim.

Grasp the victim's wrist from above, your thumb and index finger are facing you. The victim should never be allowed to grab you and put your life at risk. Therefore, never reach across the victim to grab his wrist, always grab the wrist that is the closest to you.

Keep your arm straight and locked out and pull the victim to the side of the deck. The victim should never be allowed to grab you and put your life at risk. Therefore, never pull the victim into you, always pull the victim into the side.

Arm Extension

An arm extension rescue can be used if you cannot reach the victim using the reach from a deck rescue technique and you must enter the water. An arm extension rescue technique is used for a victim who is either active or passive. Once you determine that a reach from a deck rescue technique is not viable—

Reassure the victim and quickly ease into the water while holding onto the deck with one hand.

Grasp the victim's wrist from above, your thumb and index finger are facing you. The victim should never be allowed to grab you and put your life at risk. Therefore, never reach across the victim to grab his wrist, always grab the wrist that is the closest to you.

Keep your arm straight and locked out and pull the victim to your side. The victim should never be allowed to grab you and put your life at risk. Therefore, never pull the victim into you, always pull the victim into the side.

Leg Extension

If the victim is beyond the reach of your arm, ensure that you have a firm grip on the deck, extend a leg to the victim and allow him to grab it, slowly bring the victim in closer until you can grab the victim's wrist that is holding onto your leg. Once you have a firm grasp on the victim's wrist, use the steps in the arm extension to pull the victim to safety. The leg extension rescue can only be used to rescue an active victim because the victim must be able to grab the rescuer's leg.

Wading Assist

Do not wade into water that is deeper than your chest. Talk constantly to calm the victim. If possible, do not touch the victim directly. Extend an object (e.g., stick; pack; rifle with magazine removed, chamber empty, and muzzle pointing toward victim) that the victim can grasp. Once the victim grasps the object, pull the victim slowly to safety.

Throw

If the victim is not within reach, use an expedient line to throw a lifesaving device to the victim. A lifesaving device can be any weighted item that floats (e.g., a canteen that is one fourth full of water). The lifesaving device must be secured to the end of the rope so that the rope will feed out from its coil when tossed. Talk constantly to calm the victim. Once the victim grasps the line or the lifesaving device, pull the victim toward you at a steady pace that keeps the victim's head above the water's surface. DO NOT pull so strongly that you break the victim's grip on the line. To prepare and use an expedient line and lifesaving device—

Tie a bowline at one end of the rope. (The appendix illustrates various types of knots.)

Unfasten the lid of a canteen.

Place the bowline around the neck of the canteen.

Refasten the lid so the canteen hangs from the bowline loop.

Place one end of the rope under the ball of your forward foot to secure it, either tie a knot at the end of the rope or tie an object at the end of the rope to create a block. Stand with your weight on the end of the rope. DO NOT tie the rope around your ankle.

Coil 20 to 30 yards of the rope, and hold it in a nonthrowing hand.

Place the canteen in your throwing hand.

Use an underhand throw to pitch the canteen and rope a short distance over the victim's head. Keep your nonthrowing hand open so the coil can unfold freely. The rope should trail across the victim's outstretched arms.

Retrieve the rope if the throw is inaccurate or the victim fails to grasp it. Recoil the rope as it is retrieved.

Divide the coil and throw again.

Lifesaving Approaches

Properly approaching the victim is as important as any other aspect of a rescue. Knowing the appropriate and safest method needed to approach a victim is critical in ensuring your safety and the victim's survival. Determining the victim's physical state (distressed swimmer, victim that is active and drowning, or victim that is passive and drowning) is crucial and will determine which type of approach you will execute.

Front Surface Approach

The front surface approach is typically performed when the victim is passive. You must remember that approaching the victim from the front is dangerous because a distressed swimmer or victim that is active can lunge toward you.

Rear Approach

If the victim is active, you must remember that his response can change rapidly. Therefore, approaching him from the front could be extremely dangerous. The victim could easily grab you and take you under the water during a state of panic. Therefore, use a rear approach when possible. When approaching a victim from the rear, reassure the victim until you get to within 2 to 6 yards of the victim; at that point, stop talking to the victim in order to maintain an element of surprise.

Approach Strokes

Before approaching a victim, you must evaluate the victim:

- Is the victim a distressed swimmer?
- Is the victim active and drowning?
- Is the victim passive and drowning?

Select your approach stroke based on your evaluation. Remember, the safety of the rescuer is paramount. Do not endanger yourself in an attempt to reach a drowning victim.

Crawl Stroke Approach Stroke

The crawl stroke (free style) is the fastest approach stroke. This stroke is used when the victim is passive and/or unconscious in the water. That is, the victim is face down, submerged or near the surface of the water, is not breathing, or is not moving. If any of these conditions exist, there is an obvious need to reach the victim as quickly as possible.

Execution of the crawl stroke approach stroke is the same as the crawl stroke (see pages 1-40 through 1-44), except that your head remains above the water's surface to allow for free breathing and maintaining eye contact with the victim.

Maintaining eye contact is critical. If the victim becomes submerged, you will have a better chance of locating him if you know his last location. Your distance from the victim will determine if your head remains out of the water the entire time you are executing the stroke. If you are 54 yards or less from the victim, you should maintain eye contact by keeping your head raised until you reach the victim. If you must swim more than 54 yards to reach the victim, make eye contact with the victim, place your face back in the water for four or five strokes, then raise your head again to regain eye contact with the victim (repeat these steps until you reach the victim).

Body position
The position of your body is horizontal to diagonal and influenced by the position of the head. This allows you to maintain eye contact with the victim.

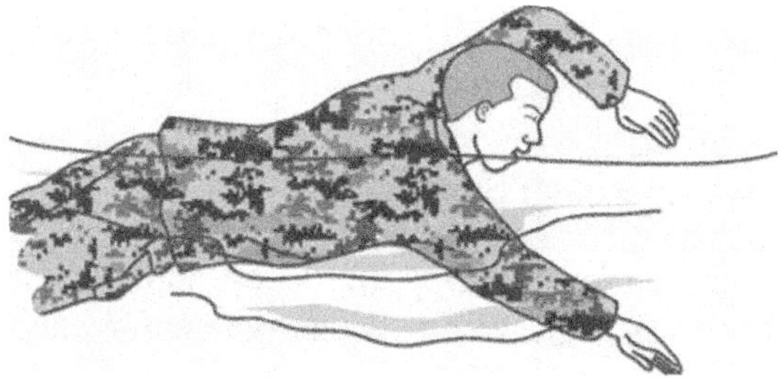

Arm action
While one arm is forward of your head to catch the water for the propulsion phase, extend your other arm alongside your body in what is known as the recovery phase. The recovery phase ends when your arm is extended forward of your head to begin the propulsion phase. In the propulsion phase, your arms are approximately shoulder-width apart and the pulling action is slightly wider and deeper to compensate for your raised head.

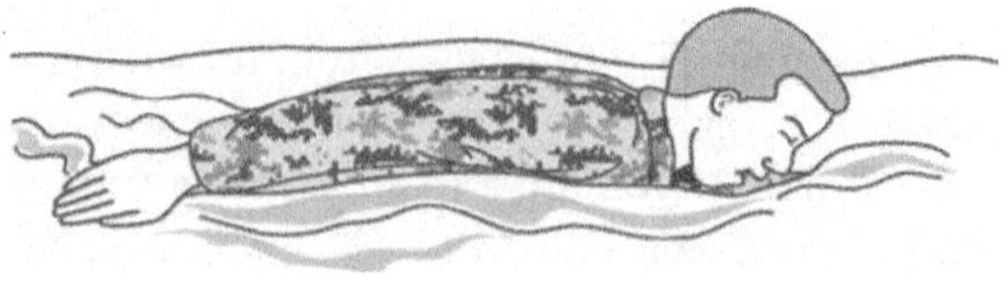

Leg action
Kick one leg in a downward motion while your other leg is recovering to the surface to prepare for the next kick (this is known as a flutter kick). This kick is used for propulsion and maintaining the lower body as horizontal to the water's surface as possible. When your head is raised, your knees are slightly bent in order to keep your feet near the surface.

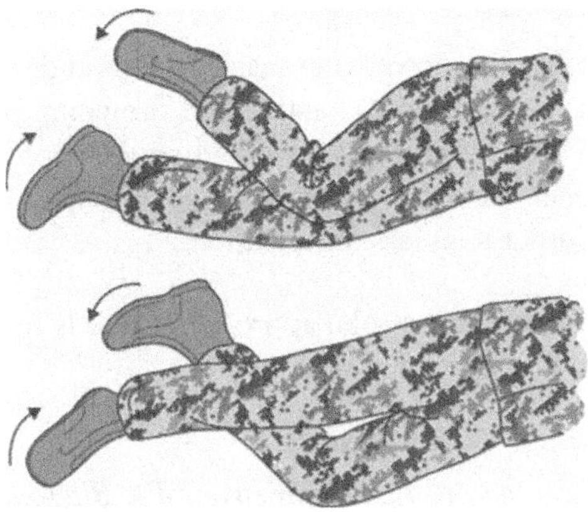

Breathing
If your face is above the water, breathe as needed (free breathing).

If traveling in excess of 54 yards to reach the victim, rotate your head to the right or left during the recovery phase and inhale when your face is out of the water or wait until your head is out of the water and facing forward while regaining sight of the victim.

Coordination
This stroke uses constant arm and leg action.

Breast Stroke Approach Stroke

The breast stroke approach stroke is used when a swimmer is distressed or a victim is active and drowning. This stroke is also used when the victim has a suspected spinal injury because it minimizes the wake created around the victim, minimizes the movement of the victim's head and neck, and helps prevent further injury.

If a victim is at a distance greater than 54 yards and if the swimmer is in distress or the victim is active and drowning in open water (ocean, lake), you should use the crawl stroke approach stroke to rapidly approach the victim, stopping approximately 11 yards from the victim in order to assess the situation.

Execution of the breast stroke approach stroke is the same as the breast stroke with a few minor deviations.

Body position
The position of your body is horizontal to diagonal and influenced by the position of the head. This allows you to maintain eye contact with the victim. It also allows you to communicate with the victim in order reassure him as you approach.

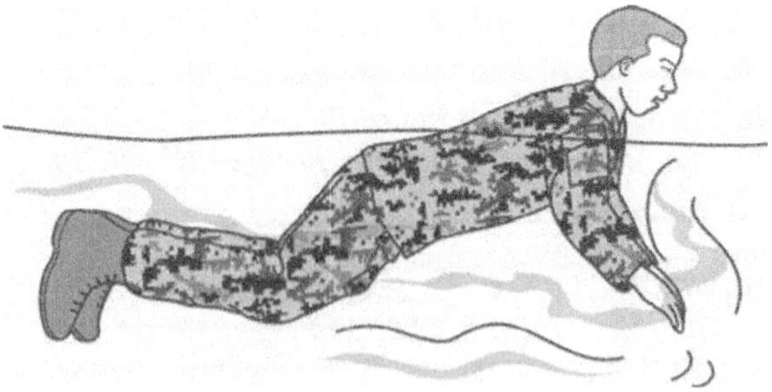

Arm action

Execute the arm action for the breast stroke, except that your arm pull is wider and deeper, which allows your head to remain above the water's surface.

Leg action

Execute the leg action for the breast stroke, except that your legs open wider during the propulsion phase of the stroke.

Breathing

Keep your face above the water and breathe as needed (free breathing).

Coordination

Coordination is the same as the breast stroke, except that there is no glide period. You constantly stroke until you reach the victim.

Level Offs

A properly performed level off positions the victim's face up and horizontal to the water's surface. There are three types of level offs: front surface approach, single armpit level off, and the double armpit level off. The single armpit level off and the double armpit level off are done by performing a rear approach, which is the preferred approach when a victim is active.

Front Surface Approach

The front surface approach is performed when the victim is passive. Approaching a victim from the front may place you in danger because a distressed swimmer or a victim who is active may lunge toward you. To execute a front surface approach—

Stop 2 to 4 yards from the victim to reassess the situation and reassure the victim.

Determine which arm to use in order to rotate the victim into a face-up position (your right hand on the victim's right wrist or your left hand on the victim's left wrist).

Move into position and prepare to grasp the victim.

Turn sideways and move toward the front of victim.

Once in position, reach forward and grab the victim's wrist. Your thumb is on the inside of the victim's wrist, as if you were checking his pulse with your thumb. Your remaining fingers wrap around the victim's wrist.

Lean back immediately and execute a powerful scissor kick or inverted scissor kick and perform short, vigorous pulls with your free arm.

As the victim begins to move forward in the water, kick, pull, and twist outboard on the victim's wrist. The momentum created from the kick and the pulling and twisting action of the victim's wrist will rotate the victim into a face-up and horizontal position in the water.

Extend and lock out your towing arm down the length of your body and execute a wrist tow to move the victim to safety.

Single Armpit Level Off

Approach the victim.

Stop 2 to 6 yards from the victim to reassess the situation, reassure the victim, and maintain your margin of safety.

Approach the victim slowly and grab his armpit with your hand (your right hand to victim's right armpit or your left hand to victim's left armpit). Position yourself sideways to the victim and place your elbow in the center of his back.

Pull with your hand that is in the victim's armpit while pushing with your elbow that is in the victim's back to place the victim horizontal on the water's surface (face up). To assist in placing the victim horizontally, use your free arm to execute short, vigorous pulls and use your legs to execute a scissor kick or inverted scissor kick.

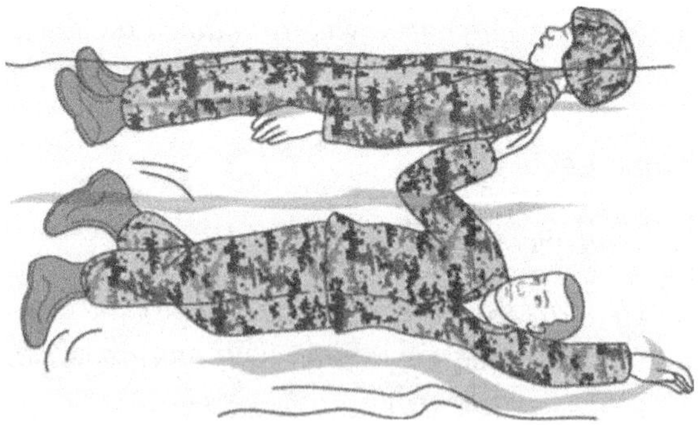

Once the victim is horizontal and on the water's surface, begin forward momentum by extending your towing arm to a fully locked-out position and executing a single armpit tow.

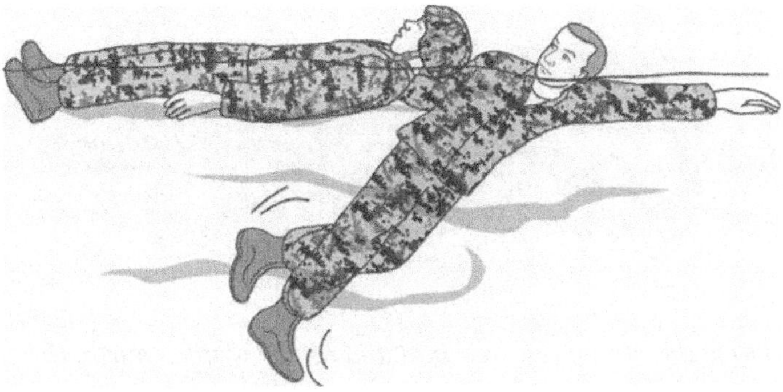

Double Armpit Level Off

Approach the victim.

Stop 2 to 6 yards from the victim to reassess the situation, reassure the victim, and maintain your margin of safety.

Approach the victim slowly and grab his armpits with both your hands. Place your elbows on his back.

Pull with your hands that are in the victim's armpits while pushing with your elbows that are in the victim's back and use an inverted breast stroke kick to place the victim horizontal on the water's surface (face up).

Once the victim is horizontal and on the water's surface, begin rearward momentum by slowly extending your arms into a fully locked-out position and executing a double armpit tow and inverted breast stroke kick.

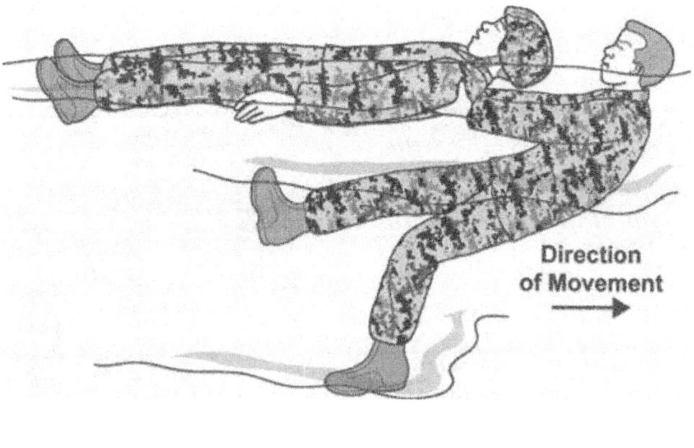

Rescue Techniques

Wrist Tow

Use the wrist tow method to rescue a victim who is floating face down. DO NOT use the wrist tow on a struggling victim. If time allows, remove your helmet and gear before attempting the rescue. Swim toward the victim using a modified breast stroke. Swim within 2 to 6 yards of the victim to maintain a margin of safety, this allows you to reassess the situation and reassure the victim. The following steps show proper front surface approach, wrist tow procedures:

Approach the victim from the front and grasp the underside of the victim's left wrist with your left hand or the right wrist with your right hand. Ensure that your thumb is on the underside of victim's wrist.

Lean back, pulling and kicking strongly to move the victim into a horizontal position.

Twist the victim's wrist to rotate the victim into a face-up position.

Swim toward safety using the lifesaving stroke.

> NOTE: The lifesaving stroke is a modified side stroke wherein the top arm is used to tow or carry a victim to safety and use a scissor kick or an inverted scissor kick is used for propulsion.

Keep a firm grip on the victim's wrist.

Keep your towing arm fully extended and along your side. This keeps the victim in the water column and prevents drag.

Ensure the victim's head does not go under water during the recovery.

Single Armpit Tow

You perform the single armpit tow after a victim has been properly leveled off. The single armpit tow uses the lifesaving stroke. To execute the single armpit tow—

Place one hand (top arm) in the victim's armpit. Your towing arm is straight and along your side.

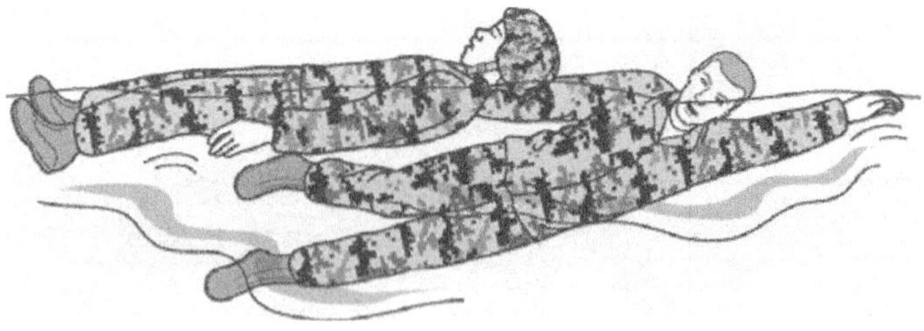

Use your lead arm (bottom arm) to execute short, vigorous pulls.

Execute a scissor or inverted scissor kick in a continuous and vigorous manner.

Use either free breathing or explosive breathing during the tow. During free breathing, keep your head above the water's surface and continuously reassure the victim. During

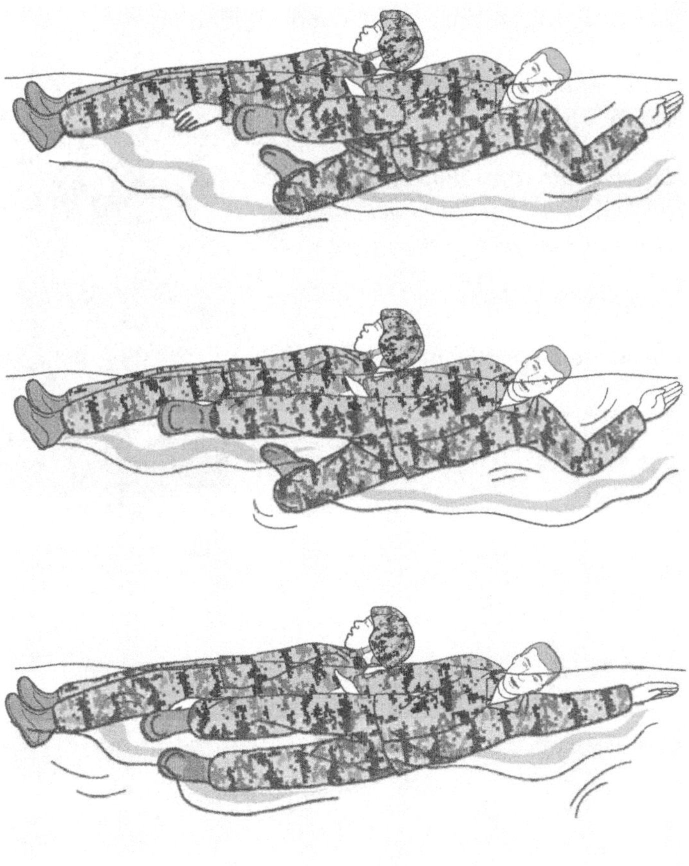

explosive breathing, put your head under the water to plane off your body's angle and reduce drag. Execute approximately two strokes, then lift your head up to breath and reassure the victim. Return your head back underwater for the next few strokes.

Double Armpit Tow

You perform the double armpit tow after a victim has been properly leveled off. To execute the double armpit tow—

Place both hands in the victim's armpits.

Extend both arms fully along your body. You are on your back, and your face is clear of the water.

Use free breathing.

Reassure the victim at all times.

Use the inverted breast stroke kick (it is the only kick that can be used due to your body position) to tow the victim to safety. (The inverted breast stroke kick is the same kick used in the elementary backstroke.) Your kick must be continuous and vigorous in order to keep the victim's face above the water.

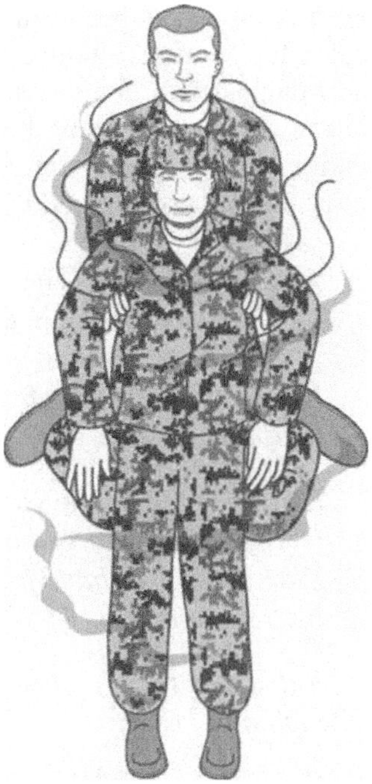

Collar Tow

You perform the collar tow after a victim has been properly leveled off. To perform a collar tow using the victim's blouse—

Maintain control of the victim by grasping his armpit with one hand and then with your free hand grasp either his combat gear or his blouse between his shoulder blades. If grasping the blouse, grasp the material with your palm up, then turn your hand over to tighten the material.

Release the victim's armpit once control is established and execute the lifesaving stroke to tow the him to safety.

Cross-Chest Carry

Use the cross-chest carry to carry a victim to safety if the victim is struggling or when moving through heavy surf. Remove your helmet and gear before attempting a rescue. Talk constantly to calm the victim. Swim toward the victim using a breast stroke approach stroke. Swim within 2 to 6 yards of the victim to maintain a margin of safety, this allows you to reassess the situation and reassure the victim. The following steps illustrate proper cross-chest carry procedures:

> **CAUTION:** The cross-chest carry causes fatigue even if you are in excellent physical condition.

Use a level-off technique to place the victim in a horizontal, face-up position.

Retain a grip on the victim with one hand. Reach over the victim with your free hand to encircle the victim's chest. Place your free hand on the victim's opposite rib cage, just below his armpit.

Release your grip once you have a secure hold on the victim's chest.

Swim toward safety using the lifesaving stroke while keeping a firm grip on the victim's chest and your hip on his back.

This procedure brings the victim's face and shoulders clear of the water, and typically the victim stops struggling. Sometimes, however, the victim will struggle during the swim to safety. If this happens, either tighten your grip on the victim or defend yourself with one of the techniques discussed on pages 2-25 through 2-30.

Tired Swimmer's Assist

You must maintain a 2 to 6 yards margin of safety from the victim at all times while you are getting into position to perform the tired swimmer's assist. Once behind the victim, you extend one arm, hand straight. Place your straight hand underneath the victim's opposite armpit. Maintain a 45 degree angle away from the victim, your arm is locked out. Assist the victim in propulsion until both of you reach safety.

Defense Against a Drowning Victim

DO NOT sacrifice your life in an attempt to save the victim. A struggling drowning victim poses great danger to anyone nearby. Driven by panic, the victim can grab you with great strength in an effort to climb out of the water. This can result in death for both you and the victim. The following techniques can be used to defend against a drowning victim, but the best defense against attack by the victim is to stay out of reach.

Block

The block prevents the victim from grabbing you if you have approached from the front. If the victim lunges toward you, react as follows:

> *Place one or both open hands against the victim's upper chest; being careful to avoid the victim's face, neck, and abdomen.*

Lean backwards and submerge rapidly. Keep your blocking arm(s) extended.

Swim underwater and away from the victim, and return quickly to the surface.

Stop 2 to 6 yards from the victim to reassess the situation.

Determine an appropriate course of action.

Wrist-Grip Escape/Wrist-Grip Escape Alternative

The wrist-grip escape is used when a victim grabs your arm or wrist. Submerge the victim quickly by reaching across with your free hand, pushing down on the victim's shoulder to submerge him, and kicking to propel yourself upward. While keeping the victim submerged with your hand on his shoulder, give three hard jerking pulls with your trapped hand in an attempt to break free from his grasp. Once free, swim clear of the victim and reassess the victim's condition.

If the wrist-grip escape is unsuccessful in freeing your hand, use the wrist-grip escape alternative: use your free hand to grab your trapped fist, rotate thumbs up, apply bone-on-bone contact with the victim's arm, pry your hand out of his grasp, and quickly swim away from the victim.

Front Head-Hold Escape

The front head-hold escape allows you to escape when you are facing a victim who is gripping you around your head and neck. To execute the front-head hold escape—

Take a quick breath and tuck your chin into your shoulder to protect your throat.

Clap your hands above your head (three times) to submerge instantly. This drags the victim below the water, lifts his arms from around your neck, and, typically, he releases his grasp in order to get back to the surface. If he doesn't release his grasp, apply pressure to the victim's brachial pressure points (which are located inside of the upper arm, above the elbow).

Thrust the victim's arms up and away.

Keep your chin tucked to protect your throat, and swim underwater away from the victim and return quickly to the surface at the ready position.

Stop 2 to 6 yards from the victim to reassess the situation.

Determine an appropriate course of action.

Rear Head-Hold Escape

The rear head-hold escape allows you to escape when a victim is gripping your head and neck from the rear. To execute the rear head-hold escape—

Take a quick breath. and tuck your chin down, turn your head to either side, and raise your shoulders to protect your throat.

Take a strong stroke, clap your hands above your head (three times), and submerge instantly. This drags the victim below the water and, typically, he releases his grasp in order to get back to the surface. If he doesn't release his grasp, apply pressure to the victim's brachial pressure points (which are located inside of the upper arm, above the elbow).

Thrust the victim's arms up and away.

Twist your head and shoulders until free.

Swim underwater away from the victim and return quickly to the surface.

Stop 2 to 6 yards from the victim to assess the situation.

Determine an appropriate course of action.

Administering First Aid/Rescue Breathing

To administer first aid/rescue breathing, remove the unconscious victim from the water when possible and stop any bleeding.

First, open the airway and check for breathing. If the victim is not breathing, give two breaths of air. Check for a pulse. If there is a pulse but the victim is not breathing, continue rescue breathing. If there is no pulse, start cardiopulmonary resuscitation (CPR). For detailed information on administering CPR and rescue breathing, refer to American Red Cross CPR skill cards.

WARNING

If the victim has no pulse and is not breathing, administer CPR immediately. If the victim does have a pulse but is not breathing, give rescue breathing only. If the victim has a pulse and is breathing, DO NOT give CPR—CPR could prove fatal.

Next, protect any wounds from exposure, and treat for shock.

Chapter 3
Treatment of Casualties and Avoidance of Dangerous Marine Life

To survive in the water, you will face many challenges other than just being able to swim. For example, injuries and/or fatigue can lead to drowning and exposure to the elements can lead to hypothermia or heat injuries and may require medical attention. Finally, your presence in the water may be seen as a threat to or food source for a variety of marine life and you may be attacked. This chapter advises you on how to minimize these risks in order to stay alive.

Drowning

Drowning is suffocation by a liquid. A drowning victim inhales water into the lungs or the throat closes by reflex so that little or no water can enter the windpipe. In either case, a victim can no longer breathe.

Symptoms

One symptom of drowning is that the victim may call for help and has an expression of dread or panic. But typically a victim that is active and drowning may not call for help because he is trying to conserve his air and will not speak.

Another symptom of drowning is when the victim thrashes at the water's surface. If the victim stops or grows calmer, he has likely been overcome by fatigue, hypothermia, or a lack of air. At this stage, the victim usually has 20 to 60 seconds before going under the water's surface.

Treatment

If the victim is not breathing, begin rescue breathing. Place the victim on his back, tilt head back to open airway, pinch the nose, and give two full breaths. If the victim does not inhale during the first two breaths, reposition his head and attempt two more breaths. Check for a pulse. If a pulse is present, but the victim is still not breathing, continue rescue breathing. If a pulse is not present, begin CPR. See MCRP 3-02G, *First Aid*, for rescue breathing and CPR details.

WARNING

If the victim has no pulse and is not breathing, administer CPR immediately. If the victim does have a pulse but is not breathing, give rescue breathing only. If the victim has a pulse and is breathing, DO NOT give CPR—CPR could prove fatal. Continue first aid until medical help arrives.

A victim who is not breathing and has no pulse may appear dead. DO NOT decide that death has occurred. Continue with the prescribed treatment. A corpsman or medical officer should decide whether the victim can or cannot be revived.

Hypothermia

Hypothermia is the abnormal lowering of the body's internal (or core) temperature to 95 degrees or below. It occurs when the body loses heat faster than it produces heat. A hypothermia victim loses the ability to move quickly, becomes mentally sluggish, slips into semiconsciousness, lapses into a coma, and dies when internal body temperatures drop too low.

The chilling effects of cold air, wind, or water can produce hypothermia. Water poses the greatest threat because it transfers heat 25 times faster than air. Depending on the water's temperature, a victim can succumb to hypothermia within a few minutes. The body's sudden contact with cold water can also set off a body reaction known as the mammalian diving reflex. This reflex can greatly increase survival time (especially for women and children) in or under cold water. The mammalian diving reflex shuts off blood circulation, except for the flow between the heart, lungs, and brain. The small amount of oxygen left in the blood and lungs is saved for the body's vital organs. This reflex has allowed people to survive being under cold water for an extended period of time. Therefore, a cold water drowning victim should be treated as if still alive even though the victim is not breathing, has no pulse, and may appear dead. DO NOT decide that death has occurred. Continue with the prescribed treatment. Victims of hypothermia can appear to be dead when they are not. A corpsman or medical officer should decide whether the victim can or cannot be revived.

WARNING

If the victim has no pulse and is not breathing, administer CPR immediately. If the victim does have a pulse but is not breathing, give rescue breathing only. If the victim has a pulse and is breathing, DO NOT give CPR—CPR could prove fatal. Continue first aid until medical help arrives. Check for a pulse for at least 45 seconds.

Symptoms

Once the body's core temperature drops, the victim will show one or more of the following symptoms:

- Violent and uncontrollable shivering as the body tries to warm itself.
- Slow or slurred speech.
- Disorientation or poor coordination.
- Loss of skin color.
- Blue and pinched lips.
- A slowing or stopping of shivering that progresses into a rigid torso and limbs.

Survival Time

A hypothermia victim's survival depends on the water's temperature and the time spent in the water. A small body build cools faster than a large build. Children cool faster than adults.

To increase your chance of survival in the water, utilize the HELP position described on page 1-37. Extra clothing and inactivity (remaining motionless in the water) can also increase your survival time.

Treatment

A hypothermia victim must be warmed to prevent further heat loss; therefore, treatment should begin as soon as possible. Consciousness of the victim determines the treatment that should be pursued. See MCRP 3-02G for specific treatment information. The following treatment procedures are recommended:

- If the victim is conscious, give the victim warm fluids. Give candy or sweetened foods to a victim who is able to eat.
- If the victim is unconscious, place him on his back with his head tilted back to ensure an open airway.
- DO NOT massage the victim. Massage can break blood vessels and create swelling, internal pressure, and blocked blood circulation.
- DO NOT give alcohol to the victim. Alcohol lowers the victim's body temperature.
- Shock is a possibility, treat accordingly.
- Seek medical help immediately.

If you are able to remove the victim from the water, apply the following steps when possible:

- Get the victim into shelter.
- Remove the victim's wet clothing.
- Put the victim in dry clothing.
- Place the victim in a sleeping bag if one is available. It may be necessary to place another Marine in the sleeping bag with the victim.
- Place as much insulation as possible between the victim and the ground.
- Use hot water bottles, electric blankets, or blankets heated in an oven or by a campfire to warm the victim's neck, groin, and the sides of the chest.

CAUTION: DO NOT apply heat to extremities.

Heat-Related Injuries

Heat-related injuries include heat cramps, heat exhaustion, and/or heat stroke. They can occur when a Marine—

- Is exposed to extreme heat, such as from the sun or a combination of high air temperatures and water temperatures.
- Does not wear proper clothing or gets overexposed to the sun.
- Becomes dehydrated.

With heat cramps, muscles may cramp in the arms, legs, and/or stomach. The victim may sweat excessively. To treat, create improvised shade for the victim and have him drink water.

With heat exhaustion, the victim sweats heavily; presents pale, moist, cool skin; and complains of a headache, weakness, dizziness, and/or loss of appetite. A victim may also experience heat cramps, nausea, vomiting, an urge to defecate, chills, rapid breathing, confusion, and a tingling sensation in the hands or feet. To treat, pour water over the victim and fan him to speed up the coolant effect of evaporation, have the victim drink freshwater, and attempt to provide shade.

With heat stroke, the victim stops sweating, which results in red, flushed, hot, dry skin. The victim may first experience headaches, dizziness, nausea, fast pulse and respiration, and/or seizures and mental confusion. The victim can become unconscious and die if not treated quickly. To treat, create improvised shade for the victim. Have him drink water and elevate both legs if possible.

Burns

Apply a field dressing (or the cleanest material available) to the burn. Give sips of water to a casualty who is conscious and not nauseated. When treating burn casualties—

- DO NOT remove clothing stuck to the burns.
- DO NOT break any blisters.
- DO NOT apply grease or ointments to the burns.

For electrical burns, check for both an entry and exit burn from the passage of electricity through the body. An exit burn may appear on any area of the body, not necessarily opposite the entry burn.

For burns caused by wet or dry chemicals, flush the burns with large amounts of water and cover with a dry dressing.

For burns caused by white phosphorus, flush the area with water, then cover with a wet material, dressing, or mud to exclude air and keep the white phosphorus particles from burning.

For laser burns, apply a field dressing.

Common Medical Problems Associated with Sea Survival

Seasickness

Seasickness is the nausea and vomiting caused by the bobbing motion created by the wave action of a flotation device. Seasickness can result in—

- Dehydration and exhaustion.
- A loss of the will to survive.
- Others becoming seasick.
- Unclean conditions.

To treat seasickness—

- Wash both the Marine and the flotation device to remove the sight and odor of vomit.
- Keep the Marine from eating food until the nausea is gone.
- Have the Marine lie down and rest.
- Give the Marine seasickness pills if available. If the Marine is unable to take them orally, the pills should be inserted rectally for absorption by the body.

Saltwater Sores

Saltwater sores occur when skin that has abrasions or is cut is exposed to saltwater. The sores may form scabs and pus. Do not open or drain. Flush the sores with freshwater, if available, and allow to dry. Apply antiseptic, if available.

Blindness/Headache

Irritants or the effects of the sun's rays reflecting off the water can cause temporary blindness or headaches. If flames, smoke, or other irritants get in your eyes, flush the eyes immediately with saltwater, then with freshwater, if available. Apply an ointment, if available. Bandage both eyes for 18 to 24 hours or longer if the damage is severe. If glare from the sky and water causes your eyes to become bloodshot and inflamed, bandage the eyes lightly. Try to prevent this problem by wearing sunglasses or goggles with a sunglass insert.

Constipation

This condition is a common problem associated with dehydration. For constipation, do not take a laxative if it is available; this causes further dehydration. Drink freshwater, if available.

Sunburn and Dehydration

The sun's rays reflect at all angles off the waves of the water; therefore, sunburn and dehydration are serious problems in sea survival. Try to prevent sunburn by—

- Erecting an improvised canopy, with available floating materials, to provide shade.
- Wearing your soft cover or using a cloth, such as a handkerchief, to cover your head.
- Covering your skin with sunscreen or lip balm from your first aid kit. Your lips, nostrils, eyelids, the backs of your ears, and the skin under your chin sunburns easily. If enough sunscreen cream is available, all exposed skin should be covered.

Dehydration is caused by the loss of the body's vital fluids. Dehydration in saltwater may result from a combination of factors such as a lack of water, the effects of saltwater on skin tissue, sunburn or vomiting from seasickness, and other causes. Sleep and rest and reduced water and food intake are the best ways of enduring periods of exposure. The following measures will delay the effects of dehydration:

- DO NOT drink saltwater.
- DO NOT drink urine.
- DO NOT drink alcohol.
- DO NOT smoke.
- DO NOT EAT unless water is available.

Dangerous Marine Life

You may see many types of marine life around you, and some are more dangerous than others. Generally, sharks are your greatest danger, followed by barracudas. However, most marine life will not deliberately attack a human. The most common injuries from marine life are wounds from bites, stings, or punctures. With the exception of sharks and barracudas, most injuries are a result of either trying to catch game or from contact abrasion with marine life. To treat an injury that results from a dangerous marine animal bite—

- Pack the wound with gauze and then apply a pressure bandage, if available.
- Treat for shock.
- Obtain medical attention as soon as possible.

For a more detailed information on the treatment of casualties and dangerous marine life, see MCRP 3-02F and MCRP 3-02G.

Sharks and Barracudas

Only about 20 percent of all shark species are known to attack people. Sharks have an acute sense of smell, and the smell of blood in the water will draw them to their prey. They are also very sensitive to any abnormal vibrations in the water; therefore, the sound caused by a struggling swimmer or underwater explosions will attract them.

Barracudas are bold and inquisitive fish. They have been known to attack men who are wearing shiny objects. Barracudas may charge at lights or shiny objects at night.

A group of swimmers can maintain a 360 degree watch while in the water. Therefore, to protect yourself from sharks and barracudas, stay with other swimmers. The group members can either frighten away or fight off sharks or barracudas better than one person can. Keep all your clothing on, including footwear. Historically, sharks attack unclothed individuals in groups first, mainly in the feet. Clothing also protects you against abrasions from a shark's tough skin should the shark brush up against you. Avoid urinating heavily, let urine dissipate between discharges. If you must defecate, do so in small amounts and toss it as far away from the group as possible. Do the same if you must vomit.

If attacked, the use of firearms by swim sentries should be used with extreme caution because of the risk of injury to other swimmers. If unarmed or unable to make an improvised weapon, kick and strike the shark. Avoid using your bare hands to strike the shark, injury can result to your hands due to a shark's tough skin. Target areas on a shark are the gills, eyes, and underbelly. Blows to the snout are also not recommended because a shark will tilt its head up and thrust its jaws forward when biting.

Sea Snakes

Sea snakes are venomous and sometimes found in mid-ocean. They are unlikely to bite unless provoked. Your best protective measure is to avoid them.

Poisonous Fish

Many reef fish are toxic, and can kill you if eaten. Generally, poisons are present in all parts of the fish, but especially concentrated in the liver, intestines, and eggs.

Turtles and Moray Eels

Turtles and moray eels normally inflict minor bite wounds. Treat this type of injury by cleaning the wound.

Corals

Coral, dead or alive, can inflict painful cuts. Clean all coral cuts thoroughly. DO NOT use iodine to disinfect any coral cuts. Some coral polyps feed on iodine and may grow inside the flesh.

Jellyfish, Portuguese Man-of-War, Anemones, and Others

This group of marine animals inflicts injury by stinging their victims with their tentacles. Contact with their tentacles produces burning pain, a rash, and small hemorrhages on the skin. Shock, muscular cramping, nausea, vomiting, and respiratory distress may also occur. Gently remove the clinging tentacles with a towel and wash or treat the area. Use diluted ammonia or alcohol and talcum powder to treat the injury if available.

Spiny Fish, Urchins, Stingrays, and Cone Shells

These animals inject their venom by puncturing the skin with their spines. General signs and symptoms include swelling, nausea, vomiting, generalized cramps, diarrhea, muscular paralysis, and shock. Deaths are rare. Treatment consists of soaking wounds in hot water, if available, to deactivate heat-sensitive toxins.

Chapter 4
Negotiating Water Obstacles

Marines face water obstacles in saltwater, freshwater, and brackish water (where freshwater and saltwater meet). These water environments differ considerably and pose distinct problems for Marine tactical units and swimmers. Saltwater obstacles include tides, surfs, and currents. Brackish water obstacles include back bays. Freshwater obstacles include rivers and canals.

Tides

Tides are periodic changes in the surface levels of oceans, bays, gulfs, inlets, and rivers. The Moon's and Sun's gravitational pulls cause tides. Depending on the situation, tides can either help or hinder Marines in their endeavors to conduct amphibious or riverine operations. Tides can provide sufficient depths to allow for the passage of landing craft or boats over reefs, trees, and other underwater obstructions. On the other hand, tides can render a river fording site unusable to tactical vehicles. Direction, level of change, and amount of change determine tidal nomenclature.

Tides that show change in direction are flood tides and ebb tides. Rising tides are known as flood tides. Falling tides are known as ebb tides.

Tides that show extreme levels of change are high tides and low tides. High tide is the period when water is at its greatest depth. Low tide is the period when water is at its most shallow depth.

Tides that show the least and most amount of change are neap tides and spring tides, respectively. Neap tides have the least amount of change in water levels between high and low tide. Neap tides occur at the half moon, when the Sun and the Moon are aligned at a 90-degree angle with the Earth. In this position, the Sun's and Moon's gravitational pulls offset each other. Spring tides have the highest floods and lowest ebbs. Spring tides occur at or shortly after the new moon or full moon when the Sun, Moon, and Earth are approximately in line. In this position, the Sun's and Moon's gravitational pulls are combined.

Surf

Waves break upon entering shallow water and create surf. The offshore area where waves break is the surf zone, which can present many hazards. Breaking waves often trap air bubbles and create a foamy appearance. Bubbles lower the water's density and decrease buoyancy. Move through foamy surf as quickly as possible. The type of wave determines your survival technique.

Wave action moves you toward shore. Lie on your back or side with your head pointing in the direction of the beach and your feet pointing into the waves.

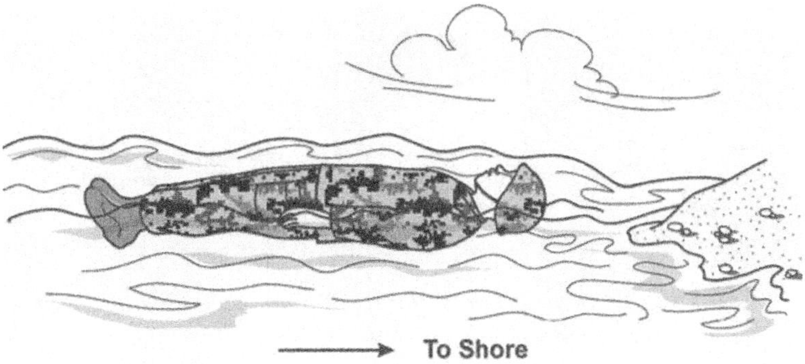

As one wave approaches the beach, another drains away; relax and do not swim against the draining water. When a new wave is within about 3 yards, start swimming toward shore. Continue to swim until the wave lifts you and moves you toward the beach. Once the wave loses forward momentum, relax and repeat the cycle. If nearing rocks, turn your body and approach feet first to reduce the chance of striking your head and arms.

Plunging Waves

A plunging wave is a breaker whose crest curls forward and falls ahead of its base. Because of its power and underwater turbulence, a plunging wave poses the greatest surf threat. If caught in a plunging wave, you can be pulled underwater and pitched about violently. This can cause you to panic, which can increase your chance of drowning. Perform the following steps to escape a plunging wave:

Tuck into a ball with your head against your knees and your forearms locked around your legs, just below the knees.

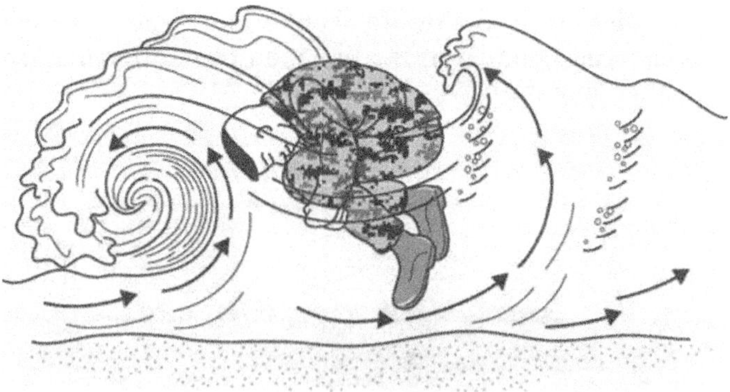

Relax in this position until the turbulence subsides and you float to the surface. This can take 30 seconds or more.

Swim toward shore.

> *Note:* If threatened by another plunging wave, dive underwater into the wave.

Spilling Waves

A spilling wave does not break. Instead, its crest slides forward without curling. A spilling wave creates less turbulence and poses less of a threat than a plunging wave. If caught in a spilling wave, relax and let the wave carry you to shore.

Surging Waves

A surging wave occurs on a beach with a steep underwater gradient. It never really breaks, but the crest rises while the base slides up the beach with great force and speed. Once the wave reaches its highest point on the beach, it rushes back as quickly as it surged forward. If you are standing on the bottom when a surging wave advances or retreats, the wave can knock you off your feet and pull you into the surf zone. Do not try to stand or walk on the bottom. Swim toward the beach as soon as possible.

Currents

Offshore Currents

An offshore current occurs outside the surf zone. Typically, it occurs at bay entrances, in island channels, and between islands and the mainland. An offshore current flows parallel to or away from shore. If the offshore current is created by tides, its current strength and direction vary at different times of the day.

If caught in an offshore current, you may be carried in a direction you do not want to go. DO NOT try to swim directly to safety. If the current is moving directly away from the shore, relax and wait until the current dies out or turns toward land. Once the current subsides, use any survival stroke to swim toward shore. If the current is moving parallel to shore, use any stroke to move at an angle across the current and toward shore.

Rip Currents

A rip current occurs when waves pile water against the shore faster than the water can drain. The water flows rapidly along the beach until it is deflected seaward by a bottom obstruction. Then the rip current flows through the surf zone and into open water at a speed of up to 2 knots. This action can cut deep trenches in the sand. A rip current dies out once it reaches open water (usually within a few hundred yards of the shore).

A rip current can pull you out to the open sea. If caught in a rip current, DO NOT try to swim against the current. A rip current moves faster than most people swim, and it is impossible to swim to shore once caught in it. Relax and stay afloat until the current runs out. Once the current subsides, use a survival stroke to move parallel to the shore until you are out of the current, then begin swimming toward shore.

Littoral Currents

A littoral current occurs when a wave breaks against a beach at an angle. This current flows parallel to the shoreline and does not pose a great threat. If caught in a littoral current, use the combat travel stroke to swim across it at an oblique angle.

Back Bays

Once on the beach, you often face one or more rows of low hills called dunes. Behind the dunes, you may encounter a low-lying stretch of ground thickly covered with scrub trees and bushes. This area gives way to wetlands known as back bays. Back bays consist of muddy islands that are almost submerged during flood tides and separated by channels of brackish water of varying depths. Channel bottoms usually contain soft mud. Back bays pose major obstacles to vehicular traffic.

Infantry can cross back bays, but only with great effort. If crossing back bays by foot, consult detailed navigation charts and use the following guidelines to plan your route:

- Avoid water less than waist deep; walking in shallow water or soft mud is extremely tiring.
- Avoid back bay islands; these low-lying islands are usually too muddy to support foot traffic.
- Seek out deep water; floating with a pack is less tiring than walking through shallow water or soft mud.
- Seek out sand, shell, gravel, or stone bottoms; these firmer bottoms generally ease travel and help conserve energy.

Rivers and Canals

A river is a large, natural stream of water that empties into a larger body of water. The slope of the riverbed and the volume of water in the river determine its current.

Canals resemble small rivers or streams in their width and depth, but usually lack any significant current. Climbing out of these waterways can be difficult if the canal is flanked by steep banks.

Chapter 5
Fording Waterways

WARNING

Fords are dangerous. Cross as quickly as possible.

A ford is any site in a river, stream, or canal where the water is shallow enough for troops or vehicles to cross without using flotation devices. Canal bottoms are usually too soft to support fording vehicles, and wading infantry frequently stumble.

The tactical situation dictates the location of the fording site. Seek fords that are protected from enemy observation and that allow for adequate supporting fires. A night fording takes at least one and a half times as long as a daylight fording.

> *NOTE:* In this chapter, the term "river" refers to rivers, streams, and canals.

Selection of a Ford Site

The following table identifies desirable fording characteristics:

Characteristics	Comments
Concealment	The ford hides personnel and vehicle movement from enemy observation.
Accessibility	The ford should have low banks with gentle gradients. This allows a free flow of traffic at both the entrance and the exit.

Characteristics (Continued)	Comments (Continued)
Slow Current	The ford's current should not exceed 1.5 meters per second if possible.
Firm Footing	The ford's bottom, entry, and exit composition should be firm enough to support traffic. Do not drive a vehicle over any bottom composition that a 2-inch diameter stick can be pressed into more than 1 or 2 inches.
Gently Sloped Channels	The ford's entry and exit points should be gently sloped. If possible, locate a portion of the stream where the channel is not actively shifting.
Depth	The fording depth is less than or equal to the least capable vehicle.

Determine the Slope

Units move into and out of water faster and more quietly if entry and exit points are not steep or muddy. Slope is the amount of change in ground horizontal distance (run) and in vertical elevation (rise) from one point to another. Slope is usually expressed as a percentage. You can use a clinometer, map, or line of sight and pace to measure the percentage of a slope.

Clinometer

A clinometer measures percentage of a slope. It is a component of the M9 armored combat excavator, and organic to many Marine engineer units. However, engineers are also taught (and often use) field-expedient methods to determine slopes and gradients.

Map

A map measures the horizontal distance along a desired path. Determine the difference in elevation between the path's starting and ending points. From a map's scale, you can determine the distance between two points. From a map's elevation lines, you can determine the difference in height between the same two points. Both figures must be in the same unit of measure (e.g., feet, meter). Divide the elevation (rise) by the distance (run) and multiply by 100.

Rise = 165 - 120 = 45 meters

Run = 200 meters

% Slope = $\frac{45}{200}$ **x 100 - 22.5%**

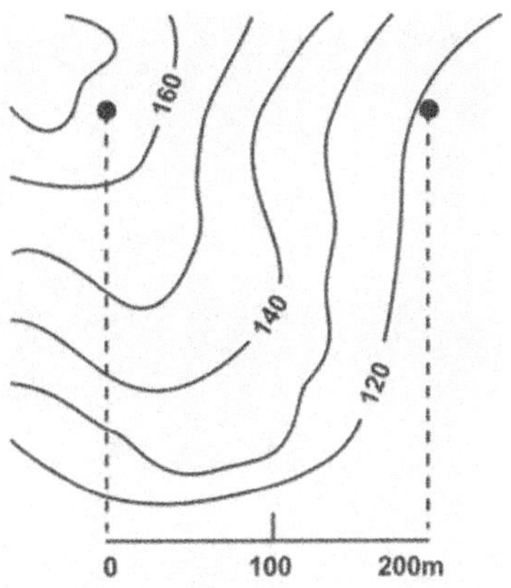

Line of Sight and Pace

To determine line of site and pace, stand at the bottom of the slope, keep your eyes level, pick a spot on the slope, then pace the distance. The number of paces multiplied by a standard measure of 0.75 meter determines the run. The eye-level height (usually 1.5 to 1.7 meters) determines the rise. Repeat this procedure until you have covered the entire distance you want to measure for each spot (vertical and horizontal). Add the vertical distances to provide total rise and the horizontal distances to provide total run.

Determine the Current Speed

Current speed increases as channels narrow. It may be necessary to locate a wider ford location to obtain a slower stream current. The following steps are used to calculate the speed of the current:

Determine points A and B along a channel. Then measure the distance between those two points.

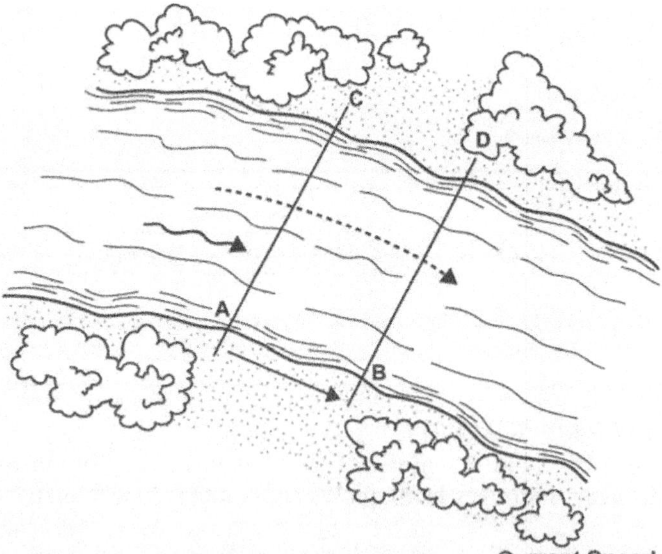

$$\frac{\text{Distance (meters) Between Points A and B}}{\text{Time (seconds) Between Points C and D}} = \text{Current Speed (meters per second)}$$

Sight directly across the water from points A and B to locate points C and D.

Throw a floating object (e.g., a stick) upstream from points A and C. Observe the object as it floats toward points B and D.

Subtract to find the time it takes for the object to float from start to finish.

Do not attempt to swim across currents that are moving faster than 1.5 meters per second. Equivalents of this speed include—

- Quick-time march rate of 120 counts per minute with one, 30-inch step at each count.
- 5 feet per second.
- 3.5 miles per hour.
- 5.5 kilometers per hour.

Measure River Width

A river's width can be estimated from the width of its symbol on a scaled topographic map. If this is not possible, use the following compass techniques:

Stand at the water line (A).

Shoot an azimuth to a point on the opposite bank (B).

Move upstream or downstream until you are at a point (C) where you can shoot an azimuth 45 degrees larger or smaller than the original azimuth.

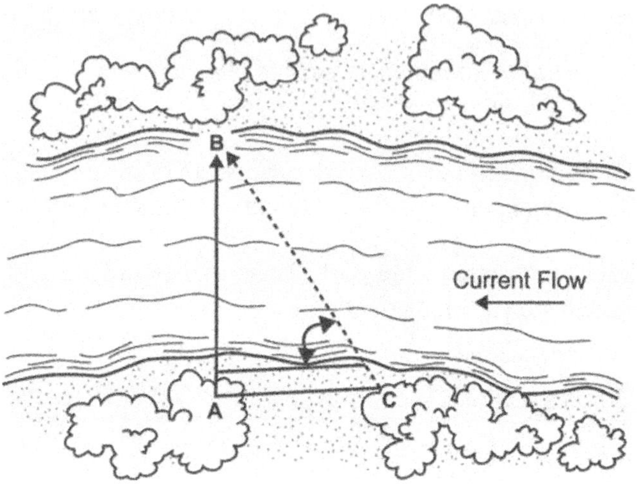

Measure the distance between points A and C. The distance calculated equals the river's width.

Calculate Downstream Drift

A river's current causes personnel and equipment to drift downstream. If personnel and equipment are aimed straight across the river, they will sideslip downstream as they move across the current to the other shore. Therefore, personnel and equipment crossings must compensate for the effects of a river's current and entries are usually made upstream of the desired exit point. Use the formula below to calculate downstream drift.

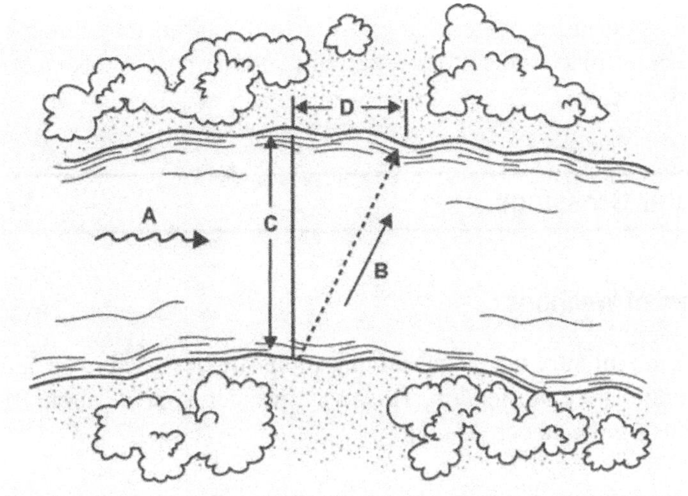

$$\frac{\text{Current Speed (A)}}{\text{Crossing Speed (B)}} \times \text{River Width (C)} = \text{Downstream Drift (D)}$$

> *Note:* The crossing speed for a swimmer across a river may vary but is generally limited to 1 meter per second. All measurements must be in the same unit of measure (e.g., meters, feet).

The Buddy System

Whenever a Marine unit must enter into or operate on the water, a "buddy system" is employed in which every Marine is paired with a swimming partner. The buddy system matches an experienced swimmer with a weak swimmer. The experienced swimmer assists and encourages the weaker swimmer and bolsters confidence during night crossings. If a unit has an odd number of Marines, place the extra person with another pair to form a three-person team.

Water Crossings

Care of Weapons

Marine infantry weapons and munitions are designed to be able to operate after immersion. However, protect your weapons from moisture whenever possible.

A gas-operated weapon can malfunction if water travels down the barrel and enters the gas tube. To protect the gas tube—

Close the weapon's bolt before entering the water.

Seal the muzzle with a condom, balloon, plastic spoon wrapper, or other form of waterproof material.

Tie or melt the protective cover to create a watertight seal.

When the muzzle's protective cover is no longer needed, remove it. Open the bolt and inspect the barrel. If the tactical situation permits, swab excess moisture from the barrel. Test fire automatic weapons, if possible. Field strip and clean weapons as soon as possible. If time does not allow for a complete inspection, rinse inaccessible areas with small amounts of diesel fuel, then dry.

ISOMAT Raft

Construction of an ISOMAT raft is time-consuming. This type of raft should not be employed as part of an attack, but used for logistical purposes (e.g., evacuating stretcher cases, transporting supplies). Use the following steps to build an ISOMAT raft:

Wrap ISOMAT sleeping pads around sturdy sticks.

Use parachute cord and square knots to tie the pads securely in place and to lash stick ends together in a rectangle.

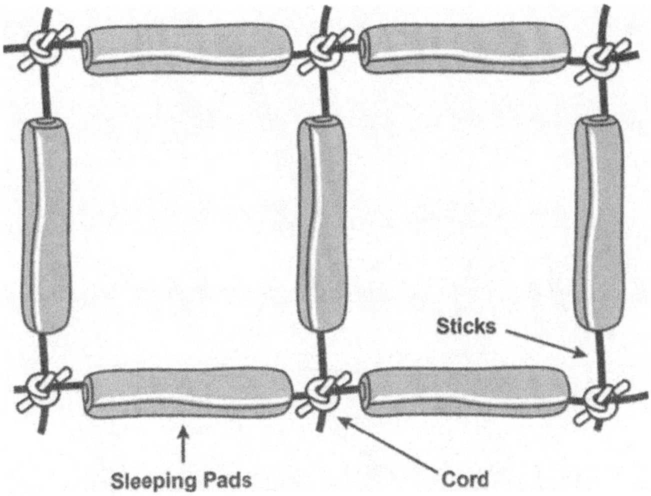

Note: The ISOMAT raft pictured can support several hundred pounds. However, the cargo will get wet if not properly waterproofed.

Poncho Raft

A poncho raft can support two Marines and their equipment and is well suited for long crossings. Use the following steps to build a poncho raft:

Inspect two ponchos and ensure they are serviceable.

Lay one poncho flat on the ground, with the hood-side up.

Cinch the hood tightly to form a gooseneck or tie in a knot.

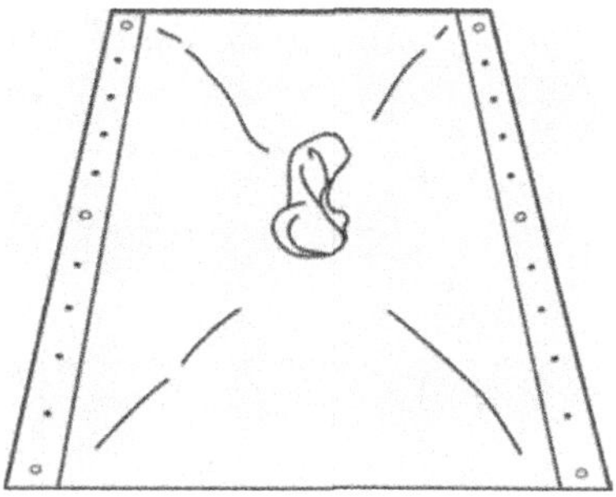

Pad sharp edges of equipment and place the equipment in the center of the poncho.

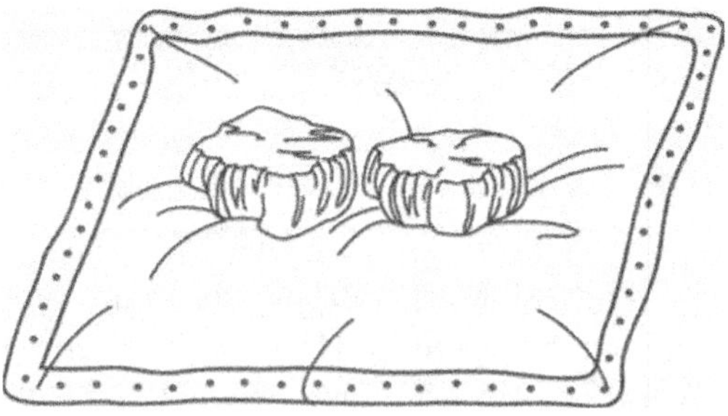

Place the second poncho over the equipment, rubber side up, and hood facing down.

Snap the edges of the two ponchos together.

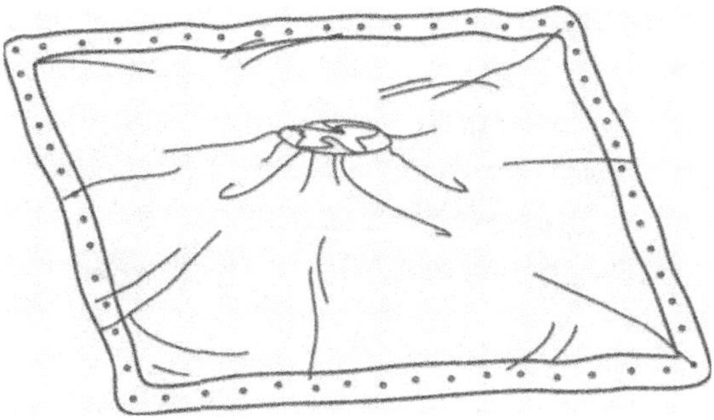

Roll the edges toward the equipment.

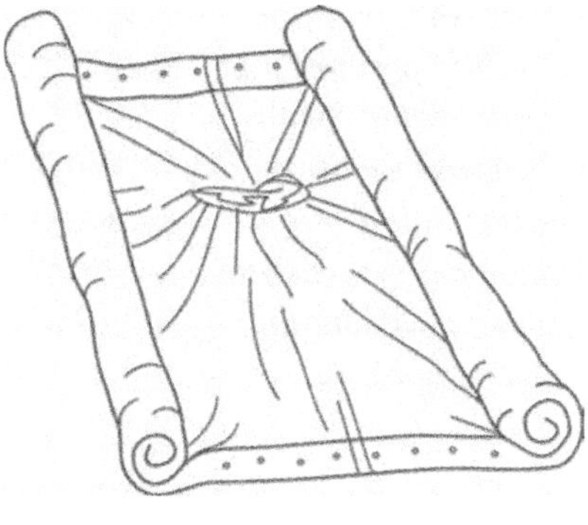

Roll the edges into pigtails and tie them off.

Pull the pigtails together over the top and lash them securely.

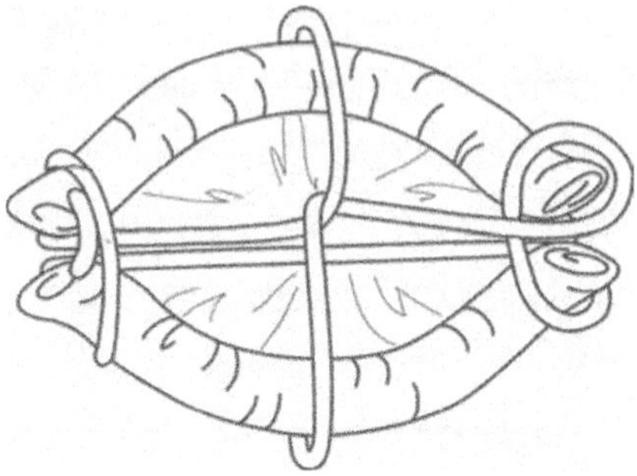

Protect the raft from brush punctures while placing it in the water. Swim across the water obstacle while security elements are covering the far shore.

Construction of a Pack Raft

You will need two waterproofed packs and two M16A2 service rifles. The following steps are required to construct a pack raft:

Place two packs side-by-side with the pack frames on the deck. The tops of the packs are opposite of each other.

Loosen the main compartment straps on both packs.

Insert one rifle on each end between the straps and the packs. The muzzles are opposite of each other. The rifles serve as one means to secure the packs together. Place the front sight post under the top flap.

Tighten the straps so that the rifles and packs are secure.

Take the excess strap on the inner side of each pack and secure it to the opposite pack to better secure the two packs together.

Take the excess straps on the outer sides of the packs and use those straps as safety lashing for the rifles.

Tuck the excess straps and check to make sure the rifles and packs are secure.

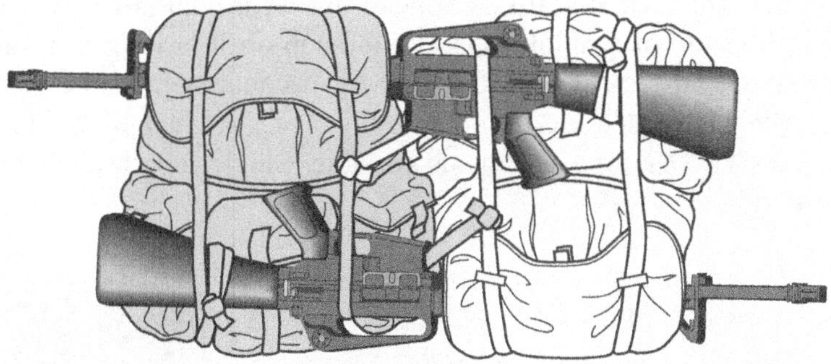

Single-Rope Bridge

A single-rope bridge offers a temporary and quick way to cross small rivers. It also provides extra security while crossing swift waters. At night, it prevents straggling, and guides units precisely from one side of the river to the other side. If crossing a river at night, plan for at least one single-rope bridge.

If your unit is crossing a river with swift currents or water depths above 4 feet, the unit is carrying sufficient rope to span the crossing site, and the tactical situation permits, secure the rope on near and far banks to provide a hand-hold for crossing Marines. This reduces the time required for the entire unit to cross and provides a degree of comfort/confidence for poor swimmers. Use a squad-sized bridge team to construct a single-rope bridge. Station several strong swimmers at the water's edge to help anyone who has trouble crossing.

Nylon rope is normally coiled in 120 foot lengths. It is 0.6 inches in diameter and has a breaking strength of about 3,840 pounds. Over time, a nylon rope can stretch to as much as one-third more than its original length and stretching weakens the rope. If the rope is stretched, discard the rope or use it for light tasks. To prolong the life of a nylon rope, do not step on it or drag it on the ground. Pad the rope in places where it contacts rocks or sharp corners. Do not leave the rope knotted or stretched longer than necessary. Dry rope as soon as possible. Single-rope bridge construction is as follows:

Tie a sling rope around your waist using a square knot and two, separate half hitches. See the appendix for detailed information on knots.

Attach a locking steel carabiner to the sling rope.

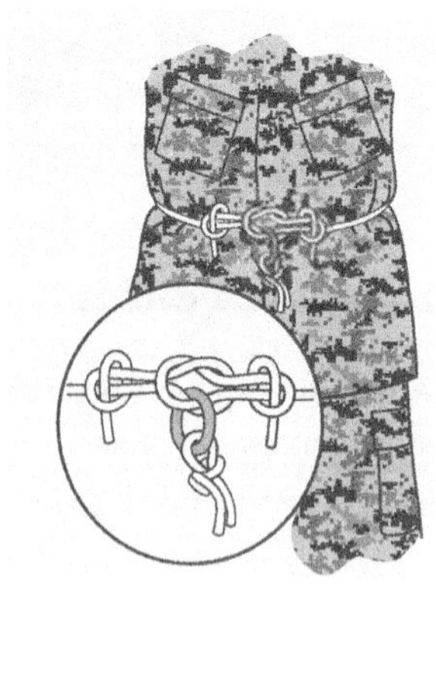

Tie a bowline knot in the running end of the bridge rope and attach it to the carabiner.

Temporarily secure the other end of the rope to a tree on the near shore.

Enter and cross the water.

NOTE: Carry only your weapon and ammunition.

Exit the water on the opposite shore.

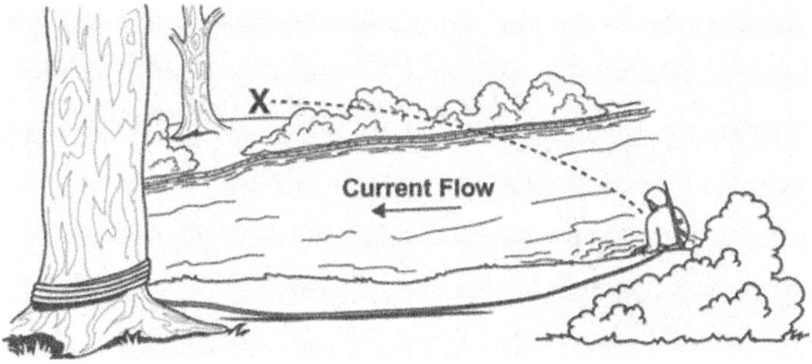

Prepare your weapon for use. Unhook the bridge rope from the carabiner at your waist, and tie the bridge rope to a sturdy tree using a round turn and two half hitches.

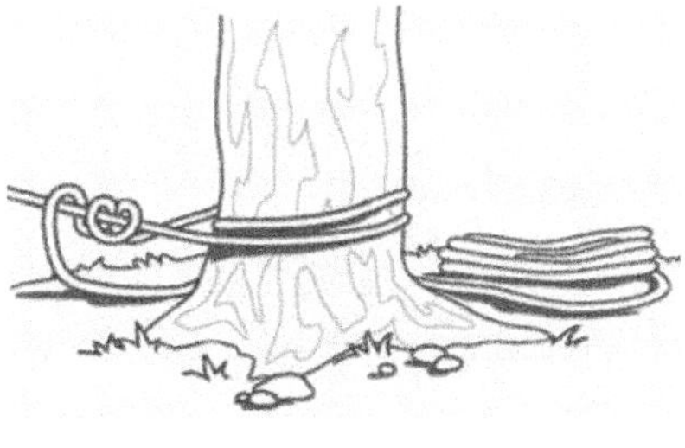

Conduct a box reconnaissance of the opposite shore.

On the near shore, have another Marine prepare to tighten the rope. That Marine should place a transport tightening system in the bridge rope by tying a double butterfly knot and placing two carabiners in the butterfly. See the appendix for detailed information on knots.

The Marine should pass the running end of the bridge rope around the downstream side of the near shore anchor point and through the two carabiners.

Pull the butterfly knot approximately one-third of the distance across the river.

Secure the bridge rope to an anchor point using a round turn and two half-hitches.

On the near shore, the Marine helping you should pull the slack out of the bridge rope until the butterfly knot is back on the near side. The bridge rope is then tied off against itself using two half hitches with a quick release in the last half hitch.

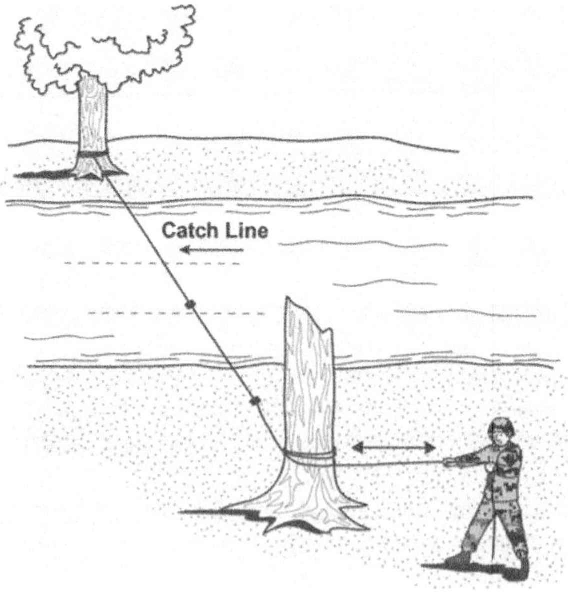

NOTE: The single-rope bridge must be as tight as possible so it will not sag when used.

If you lose your footing and fall into the water, swim with the current to the closest shore. Swimming against the current is dangerous and quickly causes fatigue.

High and Dry Crossings

If the single-rope bridge is high enough, suspend yourself below the single-rope bridge and above the water. Use the following steps to suspend yourself from a single-rope bridge and then pull yourself across the water:

Tie a sling rope around your waist using a bowline. Ensure that the knot is tight.

Attach a carabiner through the bowline's loop. The carabiner's gate faces up.

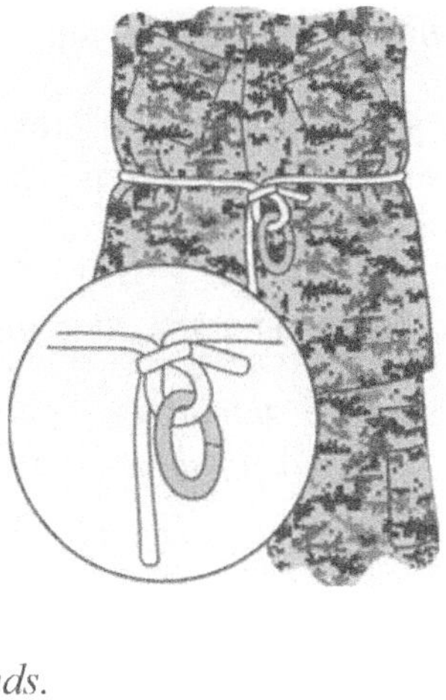

Secure your helmet chin strap.

Face the single-rope bridge with your left shoulder toward the far shore.

Grasp the bridge rope in both hands.

Swing your body beneath the single-rope bridge with your head toward the far shore. Cross your ankles above the bridge rope.

Arch back until the carabiner contacts the bridge rope. Connect the carabiner to the bridge rope. Allow the carabiner to bear your body's weight.

Pull yourself across the single-rope bridge, hand over hand, to the far shore.

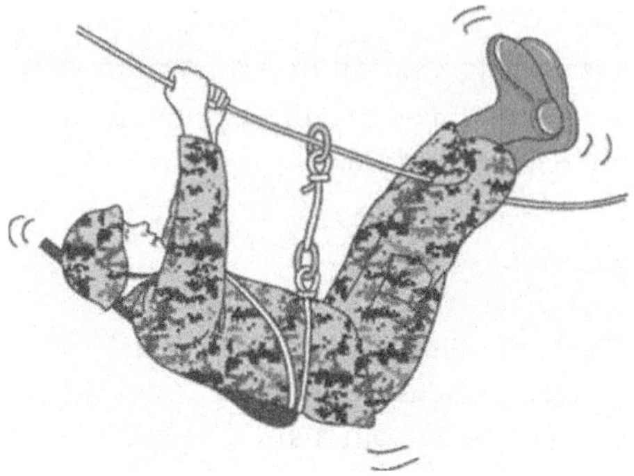

Swift Current Crossings

A single-rope bridge prevents being knocked down and swept away by a swift current. Use the following steps to move through a swift current:

Tie one end of a sling rope around your waist using a bowline.

Tie the running end of the sling rope in another bowline, and attach a carabiner to the bowline's loop.

Step up to the bridge. Face upstream.

Hook the carabiner to the single-rope bridge.

Walk sideways into the river while grasping the bridge rope in both hands.

Use the single-rope bridge for balance and remain standing, if possible.

Continue to move sideways through the river to the far shore.

Slow Current Crossings

If you face little or no current, it is not necessary to hook up to a bridge rope with a carabiner. Lie on your back in the water beneath the single-rope bridge. Support your body weight with your waterproof pack. Use the bridge rope and pull yourself hand over hand across the river.

Removal

If you are the last Marine waiting to cross, pull on the standing end of the rope to release the knot, then tie the rope around your waist using a bowline. The Marines on the far shore will pull you through the water.

Appendix
Knot Tying

Square Knot

The square knot is used to secure two ropes of equal diameter together so they form one continuous rope that will not slip.

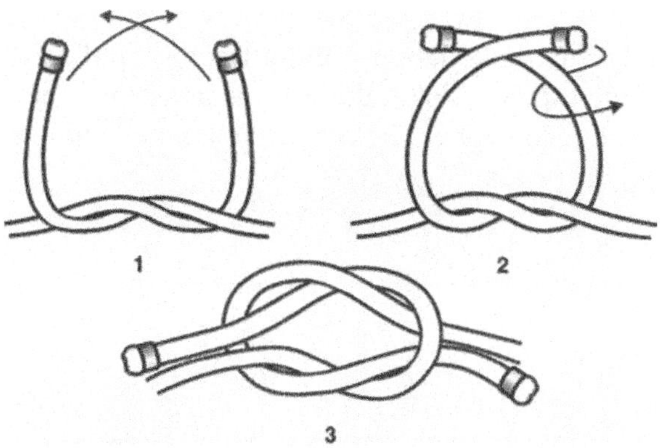

Bowline

The bowline forms a loop that will not tighten or slip under strain. It is easily untied.

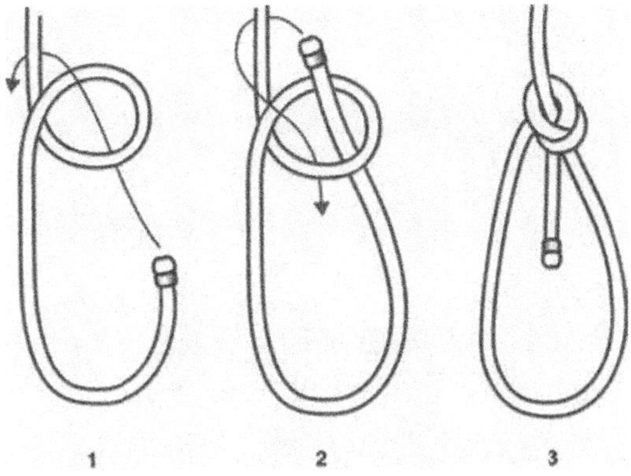

Hitches

Hitches are used to dress (prepare) knots and secure loose ends.

Half Hitch

The half hitch is used to tie a rope to a tree or to a larger rope. It will hold against a steady pull, but is not a secure hitch. It is frequently used to secure the free end of a rope.

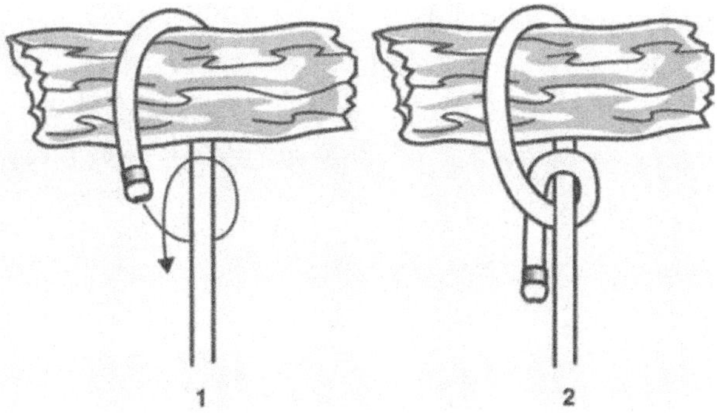

Two Half Hitches

Two half hitches can be used to secure the running end of a rope.

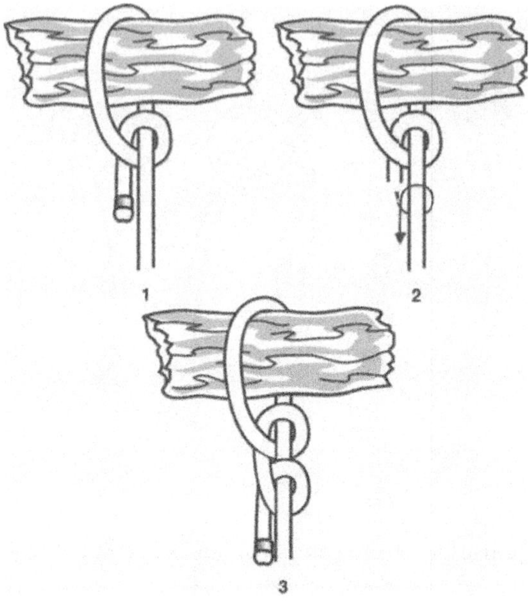

Round Turn and Two Half Hitches

A round turn and two half hitches can be used to fasten a rope to a tree. This hitch does not jam.

Butterfly Knot

The butterfly knot is used for anchor lines. Tied properly, it will not tighten on itself to the point that it cannot be easily untied. The butterfly knot can be used to pull a rope bridge taut. This knot can be used to tighten a fixed rope when mechanical means are not available. It will not jam if a stick is placed between the two upper loops.

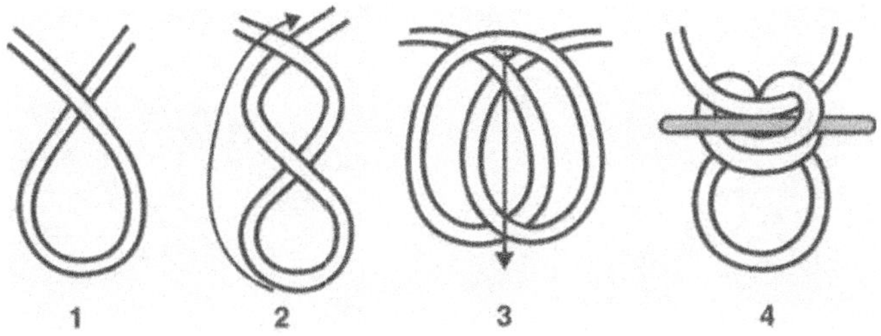

PREFACE

Combat Stress is the mental, emotional or physical tension, strain, or distress resulting from exposure to combat and combat-related conditions. Controlling combat stress is a command responsibility. In terms of Service members lost from action and reduced performance, combat stress seriously affects mission accomplishment. It is a leader's responsibility to take action to strengthen Service members' tolerance to combat stress and manage it in his or her unit.

Combat stress reactions are the result of exposure to the same conditions during military actions that cause physical injury and disease in battle or its immediate aftermath, and many combat stress reactions occur in persons who are also wounded or ill with disease. Rates of combat stress casualties vary greatly, with higher ratios during lengthy periods of intense combat. In Okinawa 1945, during a peak month of battle, the combat stress casualties among Marine Forces were reported as high as one for every two wounded in action (WIA). Under less lengthy periods, as suggested by data acquired from the Israeli Defense Forces fighting in Lebanon 1982, the ratio of combat stress casualties to WIA in small units can be as high as one to one. In the past, we have generally suffered as many as one battle stress casualty for every three to five WIA in heavy fighting. However, highly trained units with strong leadership and high esprit de corps have fewer combat stress casualties.

While this manual focuses on combat-induced stress reactions, it is important to emphasize that "combat stress" is not restricted only to combat, but may also arise from combat-like conditions present during military operations other than war. In an area of operations characterized by continuous action and high danger, our forces may experience high rates of stress casualties unless small-unit leaders are trained and prepared to manage stress.

This publication is written to inform small-unit leaders of stress characteristics and management techniques in order to *prevent, reduce, identify,* and *treat* combat stress reactions in the Service member's own unit to the maximum extent possible. A significant part of training is learning to control and cope with stress. Leaders must learn to cope with their own stress and then assist junior personnel in managing their stress. The application of combat stress management techniques helps conserve fighting strength and provides one more step toward achieving success.

Table of Contents

Chapter 2. Combat Stress Prevention, Management, and Control

Chapter 3. Command Leadership Actions

Chapter 4. Sleep Deprivation

Appendices

Chapter 1

Combat Stress Identification

1001. INTRODUCTION

The Marine Corps' success as a fighting force is dependent on leadership that maintains a balanced focus between mission accomplishment and troop welfare. The small-unit leader is the key to building and maintaining high unit morale and peak efficiency. He achieves this in part by knowing his troops and understanding their strengths and weaknesses. To maintain that same level of morale and efficiency in combat, the small-unit leader must understand how to recognize, prevent, and even personally contend with reactions to combat stress when it occurs in his unit. If a condition accounted for as many casualties in combat and the condition was at least partially preventable, the prudent combat leader would be interested in knowing more about it. Combat stress reaction(s), also called battle fatigue, is that condition. It has the potential to disable the most courageous Service member and influence the success or failure of a unit in accomplishing its mission.

1002. HISTORY

During the 1942-45 period in the European Theater, there was a ratio of one combat stress casualty for every three WIA. In a month of especially horrible, continuous fighting in Okinawa in 1945, the 6th Marine division had one stress casualty for every 1.8 WIA. However, the airborne divisions in Europe never had more than one for ten WIA, and usually less, even though they experienced very high casualties in some battles.

To accomplish a mission successfully, planners must use some guidelines to estimate losses from combat stress. What are the reasons for the tremendous range between high and low battle stress casualty rates? While the answer to this question is complex, it is clear that better-trained troops have fewer killed and wounded, and proportionally fewer stress casualties. Service members—especially leaders—can learn to recognize the symptoms and prevent or reduce the disruptive effects of combat stress.

1003. REACTIONS TO COMBAT STRESS

Service members exposed to danger experience physical and emotional reactions that are not present under more tranquil circumstances. Some reactions sharpen abilities to survive and win; other reactions may produce disruptive behaviors and threaten individual and unit safety. These adverse behaviors are collectively called *combat stress reaction*. The operative word is "behaviors." People in combat experience a range of emotions, but their behavior influences immediate safety and mission success.

Combat and combat-related military missions can also impose combinations of heavy physical work; sleep loss; dehydration; poor nutrition; severe noise, vibration, and blast; exposure to heat, cold or wetness; poor hygiene facilities; and perhaps exposure to infectious diseases, toxic fumes or substances. These, in combination with other influences—such as concerns about problems back home—affect the ability to cope with the perception of danger, and diminish the skills needed to accomplish the mission. Environmental stressors often play an important part in causing the adverse or disruptive combat stress reaction behaviors. The leader must work to keep each Service member's perception of danger balanced by the sense that the unit has the means to prevail over it. The leader must keep himself and his unit working at the level of stress that sustains performance and confidence. When troops begin to lose confidence in themselves and their leader, adverse stress reactions are most likely to occur.

It is important for the small-unit leader to recognize these adverse behaviors at the onset in order to intervene promptly for the safety and benefit of individual Service members and the unit. These behaviors may take many forms and can range from subtle to dramatic. Any Service member who shows persistent, progressive behavior that deviates from his baseline behavior may be demonstrating the early warning signs and symptoms of a combat stress reaction. Trying to memorize every possible sign and symptom is less useful to prompt diagnosis than to keep one simple rule in mind: *Know your troops, and be alert for any sudden, persistent or progressive change in their behavior that threatens the functioning and safety of your unit.*

1004. OBSERVING AND RECOGNIZING COMMON REACTIONS TO COMBAT STRESS

Ranges of fatigue, fear, anxiety, and depression affect most Service members in combat and in some military operations other than war. Mild stress reaction may be signaled by changes in behavior and only be discernible by the person himself or by close comrades. The unit leader and medical personnel depend on information from the Service member or his comrades for early recognition of combat stress reactions to provide prompt and appropriate help. Table 1-1 lists some mild stress reactions.

Table 1-1. Mild Stress Reactions.

PHYSICAL	EMOTIONAL
Trembling	Anxiety, indecisiveness
Jumpiness	Irritability, complaining
Cold sweats, dry mouth	Forgetfulness, inability to concentrate
Insomnia	Nightmares
Pounding heart	Easily startled by noise, movement, and light
Dizziness	Tears, crying
Nausea, vomiting or diarrhea	Anger, loss of confidence in self and unit
Fatigue	
"Thousand-yard" stare	
Difficulty thinking, speaking, and communicating	

Severe stress reactions may prevent the individual from performing his duties or create a concern for personal safety or the safety of others. A variety of more serious reactions or warning signs are listed in Table 1-2. These do not necessarily mean that the person must be relieved from duty, but warrant immediate evaluation and help by the small-unit leader.

Table 1-2. Severe Stress Reactions.

PHYSICAL	EMOTIONAL
Constantly moves around	Talks rapidly and/or inappropriately
Flinches or ducks at sudden sound and movement	Argumentative; acts recklessly
Shakes, trembles	Indifferent to danger
Cannot use part of body (hand, arm leg) for no apparent physical reason	Memory loss
Inability to see, hear or feel	Stutters severely, mumbles or cannot speak at all
Is physically exhausted; cries	Insomnia; severe nightmares
Freezes under fire or it totally immobile	Sees or hears things that do not exist
States vacantly, staggers or sways when standing	Has rapid emotional shifts
Panics, runs under fire	Socially withdrawn
	Apathetic
	Hysterical outbursts
	Frantic or strange behavior

Fatigue

The most common stress reactions include:

- Slow reaction time.
- Difficulty sorting out priorities.
- Difficulty starting routine tasks.
- Excessive concern with seemingly minor issues.
- Indecision, difficulty-focusing attention.
- A tendency to do familiar tasks and preoccupation with familiar details. These reactions may reach a point where the person becomes very passive, or wanders aimlessly.
- Loss of initiative with fatigue and exhaustion.

Muscular Tension

Headaches, Backache

Muscular tension increases strain on the scalp and spine and often leads to headaches, pain, and cramps.

Inability to Relax

Prolonged muscular tension wastes energy and leads to fatigue and exhaustion. Muscles must relax periodically to enable free blood flow, waste product flushing, and nutrient replenishment.

Shaking and Tremors

Mild Shaking

During incoming rounds, the individual may experience mild shaking. This symptom appears and disappears rapidly and is considered a normal reaction to conditions of great danger.

Marked or Violent Shaking

A common post-battle reaction, marked or violent shaking can be incapacitating if it occurs during the action. If shaking persists long after the precipitating stimulus ceases, or if there was no stimulus, the individual should be checked by medical personnel.

Perspiration

It is normal to experience either mild or heavy sweating or sensations of chilliness under combat stress.

Digestive and Urinary Systems

Nausea and Vomiting

"Butterflies in the stomach" is a common stress feeling. Vomiting may occur as a result of an extreme experience like that of a firefight, shelling, etc., or in anticipation of danger.

Loss of Appetite

Appetite loss may result as a reaction to stress. It becomes a significant problem if rapid weight loss occurs, or the person does not eat a sufficiently balanced diet to keep his muscles and brain supplied for sustained operations.

Abdominal Distress

Acute abdominal pain ("knotted stomach," "heartburn") may occur during combat. Persistent and severe abdominal pain is a disruptive reaction, and may indicate a medical condition.

Frequent Urination

Frequent urination may occur, especially at night.

Incontinence

During extremely dangerous moments, inability to control bowel and/or bladder functions may occur. Incontinence is embarrassing, but it is not abnormal under these circumstances.

Circulatory and Respiratory Systems

Heart Palpitations

Rapid heartbeat, a sense of pressure in the chest, occasional skipped beats, and sometimes chest pains are common with anxiety or fear. Very irregular heartbeats need to be checked by medical personnel.

Hyperventilation

Hyperventilation is identified by rapid respiration, shortness of breath, dizziness, and a sense of choking. It is often accompanied with tingling and cramping of fingers and toes. Simple solutions are increased exercise and breathing with a paper bag over the nose and mouth; or breathing slowly using abdominal muscles (called "abdominal breathing").

Faintness and Giddiness

These reactions occur in tandem with generalized muscular weakness, lack of energy, physical fatigue, and extreme stress. Brief rest should be arranged, if possible.

Sleep Disturbance

Difficulty Falling Asleep

Sometimes a Service member who has experienced intense battle conditions often cannot fall asleep even when the situation permits, or, when he does fall asleep, he frequently wakes up and has difficulty getting back to sleep.

Nightmares

Terror dreams, battle dreams, and nightmares of other kinds cause difficulty in staying asleep. Sleep disturbances in the form of dreams are part of the coping process. This process of *working through* combat experiences is a means of increasing the level of tolerance of combat stress. The individual may have battle-related nightmares or dreams that a close relative (spouse, parent) or another person important in his life has been killed in the battle. As time passes, the nightmares usually occur with less intensity and less frequency. In some cases, a Service member, even when awake, may re-experience the memory of the stressful experience as if it were recurring (called a "flashback"). This is usually triggered by a smell, sound or sight, and is not harmful as long as the Service member realizes it is only a memory and does not react inappropriately or feel overwhelmed. However, if it happens frequently or is very distressing, help should be sought from the chaplain or medical.

Restless Sleep

When a person is asleep, the sleep is not restful sleep if the person is constantly being half-wakened by noise, movement, or other stimuli. Heavy snoring often indicates poor quality sleep. The individual wakes up as tired as when he went to sleep. Finding a more comfortable position, away from distractions, can help.

Excessive Sleep

Individuals exhibiting a need for excessive sleep may be exhibiting symptoms of combat stress; however, excessive sleep is also a sign of substance abuse or depression (Persistent insomnia is a more common indicator of possible depression.)

Visual/Hearing Problems and Partial Paralysis

Stress-related blindness, deafness, loss of other sensations, and partial paralysis are not true physical injuries, but physical symptoms that unconsciously enable the individual to escape or avoid a seemingly intolerably stressful situation. These symptoms can quickly improve with reassurance and encouragement from comrades, unit medical personnel or battalion physician. If they persist, the physician must examine the individual to be sure there is not a physical cause; for example, laser range finders can cause temporary or partial blindness, and nearby explosions can cause ear damage. Individuals with these physical conditions are unaware of the causative relationship with their inability to cope with stress. They are honestly concerned with their physical symptoms and want to get better. They are willing to discuss them, and do not mind being examined. This is contrary to "malingerers" faking a physical illness, who are often reluctant to talk, or over-dramatize their disability and refuse an examination.

Visual problems include blurred vision, double vision, difficulty in focusing, or total blindness. Hearing problems include inability to hear orders and/or nearby conversations or complete deafness occurs. Paralysis or loss of sensation is usually confined to one arm or leg. Prickling sensations or rigidity of the larger joints occur. However, temporary complete immobility (with normal breathing and reflexes) can occur. If these reactions do not recover quickly with immediate reassurance, care must be taken in moving the casualty to medical for an evaluation to avoid making a possible nerve or spinal cord injury worse.

Bodily Arousal

Not all emotional reactions to stress are necessarily negative. For example, the body may become aroused to a higher degree of awareness and sensitivity.

Threat

In response to threat, the brain sends out chemicals arousing the various body systems. The body is ready to fight or take flight. The alerting systems of the experienced combat veteran become finely tuned, so that he may ignore loud stimuli that pose no danger (such as the firing of nearby friendly artillery). However, he may awaken from sleep at the sound of an enemy mortar being fired and take cover before the round hits. The senses of vision and smell can also become very sensitive to warning stimuli. The Service member may instantly focus and be ready to react.

Hyperalert

Hyperalert refers to being distracted by any external stimuli that might signal danger, and overreacting to things that are, in fact, safe. The hyperalert Service member is not truly in tune with his environment, but is "on a hair trigger." He is likely to over-react, misinterpret reassuring information as threats, and react without adequate critical thinking. Consequences can range from firing at an innocent noise to designating as hostile an innocent target.

Startle Reactions

Startle reactions are part of an increased sensitivity to minor external stimuli (on-guard reactions). Leaping, jumping, cringing, jerking, or other forms of involuntary self-protective motor responses to sudden noises are noted. The noises are not necessarily very loud. Sudden noise, movement, and light cause startle reactions; for example, unexpected movement of an animal (or person) precipitates weapon firing.

Anxiety (Fear of Death, Pain, and Injury)

Fear of death, pain, and injury causes anxiety reactions. After witnessing the loss of a comrade in combat, a Service member may lose self-confidence and feel overly vulnerable or incapable. The death of a buddy leads to serious loss of emotional support. Feelings of "survivor guilt" are common. The survivors each brood silently, second-guessing what they think they might have done differently to prevent the loss. While the Service member feels glad he survived, he also feels guilty about having such feelings. Understanding support and open grieving shared within the unit can help alleviate this.

Irritability

Reaction

Mild irritable reactions range from angry looks to a few sharp words, but can progress to more serious acts of violence. Mild irritability is exhibited by sharp, verbal overreaction to normal, everyday comments or incidents; flare-ups involving profanity; and crying in response to relatively slight frustrations.

Explosion of Aggressive Behavior

Sporadic and unpredictable explosions of aggressive behavior (violence) can occur with little or no provocation. For example, a

Service member tries to pick a fight with another Service member. The provocation may be a noise, such as the closing of a window, an accidental bumping or normal, verbal interaction.

Short Attention Span

Persons under stress have short attention spans. They find it difficult to concentrate. Short attention span causes a Service member to have difficulty following orders. The Service member does not easily understand what others are saying. The person has difficulty following directions, aiding others or performing unfamiliar tasks.

Depression

People respond to stress with protective defensive reactions against painful perceptions. Emotional dulling or numbing of normal responsiveness is a result. The reactions are easily observed changes from the individual's usual self.

Low Energy Level

The observer notes decreased effectiveness on the job, decreased ability to think clearly, excessive sleeping or difficulty falling asleep, and chronic tiredness. Emotions such as pride, shame, hope, grief, and gratitude no longer matter to the person.

Social Withdrawal

A Service member is less talkative than usual, shows limited response to jokes or cries. He is unable to enjoy relaxation and companionship, even when the tactical situation permits.

Change in Outward Appearance

If the Service member is in a depressed mood, he may be observed to exhibit very little body movement and an almost expressionless (mask-like) face.

Substance Abuse

Some Service members may attempt to use substances such as alcohol or other drugs as a means of escaping combat stress. The use of substances in a combat area makes some Service members less capable of functioning on the job. They are less able to adapt to the tremendous demands placed on them in combat.

Loss of Adaptability

Less common reactions include uncontrolled emotional outbursts such as crying, yelling, or laughing. Some Service members may become withdrawn, silent, and try to isolate themselves. Uncontrolled reactions can appear singly or in combination with a number of other symptoms. In this state, the individual may become restless, unable to keep still, and move aimlessly about. He may feel rage or fear, which he demonstrates by aggressive acts, angry outbursts or irritability.

Disruptive Reactions

Service members with disruptive, combat stress reactions—

- **Cannot function on the job.** In some cases, stress produces symptoms often associated with head injuries. For example, the person may appear dazed and may wander around aimlessly. He may appear confused and disoriented, and exhibit either a complete or partial memory loss. Service members exhibiting this behavior should be removed from duties until the cause for this behavior can be determined.

- **Compromise their own safety.** In a desperate attempt to escape the danger that has overwhelmed him, an individual may panic and become confused. The term *panic run* refers to a person rushing about without self-control. In combat, such a Service member easily compromises his safety and could possibly get killed. His mental ability becomes impaired to the degree that he cannot think clearly or follow simple commands. He stands up in a firefight because his judgment is clouded and he cannot understand the likely consequences of his behavior. He loses his ability to move and seems paralyzed.
- **Compromise the safety of others.** If panic is not quelled early, it can easily spread to others. A person in panic is virtually out of control and needs to be protected from himself. More than one person may be needed to exert control over the individual experiencing panic. However, it is also important to avoid threatening actions, such as striking him.

Referral to the Chaplain, Medical Officer or Combat Stress Control Team

Although the more serious or warning behaviors described in the preceding paragraphs usually diminish with help from comrades and small unit leaders, and time, some do not. An individual usually improves when able to get warm food, rest and an opportunity to share his feelings with comrades or small unit leader. If the symptoms endanger the individual, others or the mission, or if they do not improve within a day or two, or seem to worsen, get the individual to talk with the unit chaplain or medical officer. Access to mental health/combat stress control specialists may be sought, if available. **Do not** wait too long to see if the experience is better with time. Specialized training is not required to recognize severe stress reactions. The small-unit leader can usually determine if the individual is not performing his duties normally, not taking care of himself, behaving in an unusual fashion, or acting out of character.

1005. SUICIDE AWARENESS

Some behaviors and symptoms previously described are not only signs of stress reaction, but can also signal potential suicide risks. Service members must be ever vigilant for the signs and signals of a potential threat of suicide given by their fellow Service members. Appendix A provides useful tools for identifying these signals.

Chapter 2

Combat Stress Prevention, Management, and Control

2001. PREVENTING AND MANAGING COMBAT STRESS

The same leadership skills that apply to troop welfare and warfighting can effectively reduce or prevent combat stress reactions. Leaders should take preventive actions and address stress symptoms as they appear. Ignoring the early warning signs can increase the severity of stress reactions. Positive action to reduce combat stress also helps Service members cope with normal, everyday situations and makes them less likely to experience harmful combat stress reactions. Table 2-1 lists stress management techniques.

2002. STRESS-REDUCTION TECHNIQUES FOR LEADERS

To reduce stress, the leader should—

- Lead by inspiration, not fear or intimidation.
- Initiate and support stress management programs.
- Provide information to focus stress positively.
- Ensure each Service member has mastered at least two stress coping (relaxation) techniques, a slow one for deep relaxation and a quick one for on the job.

Table 2-1. Stress Management Techniques.

Assure every effort is made to provide for the troops' welfare.
Instill confidence in each Service member and his equipment, unit, and leadership.
Be decisive and assertive; demonstrate competence and fair leadership.
Provide sleep and/or rest, especially during continuous operations, whenever possible.
Ensure sleep for decisionmaking personnel.
Set realistic goals for progressive development of the individual and team.
Systematically test the achievement of these goals.
Recognize that battle duration and intensity increase stress.
Be aware of environmental stressors such as light level, temperature, and precipitation.
Recognize that individuals and units react differently to the same stress.
Learn the signs of stress in yourself and others.
Recognize that fear is a normal part of combat stress.
Rest minor stress casualties briefly, keeping them with their unit.
Be aware of background stress sources prior to combat; e.g., family concerns and/or separation, economic problems.
Provide an upward, downward, and lateral information flow to minimize stress due to a lack of communication.

Table 2-1. Stress Management Techniques (Continued).

Practice stress control through cross-training, task allocation, tasks matching, and task sharing.
Look for stress signs and a decreased ability to tolerate stress.
Practice and master stress-coping techniques.
Face combat stress; it is unhealthy to deny the stresses of combat.

- Look out for Service members' welfare.
- Communicate with Service members personally to learn of stressors and detect signs of stress.
- Understand that stress in response to threatening or uncertain situations is normal.
- Create a spirit to win under stress.
- Act as role model for self-control of stress reactions.

Realistic training is the primary stress-reduction technique. It assures Service members' maximum confidence in their skills and belief that their leaders are doing their best for them. Since the basic necessities of life assume even greater importance on the battlefield, leaders should:

- Ensure personnel are properly trained.
- Ensure training includes understanding of combat stress and how to deal with it.
- Place welfare of subordinates before personal welfare, but keep them capable.
- Ensure Service members get as much rest as possible.

- Ensure the best possible shelters are available.
- Keep Service members well supplied with food, water, and other essentials.
- Provide mail, news, and information avenues.
- Provide the best medical, logistical, and other support.
- Maintain high morale, unit identity, and esprit de corps.
- Keep the same unit members together.
- Assure that experienced unit members take care of and teach new members.

The unit's encouragement and support of Service members' efforts to cope with stress have a decisive effect. Unit actions can determine if Service members will endure combat exposure and accomplish their mission. Stress-coping indoctrination is part of every unit's combat training and is followed by a program of action. Controlling stress requires practice. Programs are tailored to individual units and improve with experience. Coping with stress is practiced under conditions as similar to combat as possible. A unit's ability to cope with stress in combat depends primarily on how rigorously and realistically training has been conducted.

2003. PERFORMANCE DEGRADATION PREVENTIVE MEASURES

Every Service member, team, and unit must learn to effectively sustain performance in continuous operations. This requirement applies especially to leaders. While it is an important ingredient, the determination to endure does not ensure effectiveness. Gaining the required capability goes beyond a high level of proficiency in combat skills and technical specialties. It means learning to identify the adverse conditions of continuous operations, cope with them, and overcome their effects. It also means learning how to slow the rate of performance degradation. Units, leaders, and personnel must prepare and execute plans and train to sustain performance. Adverse conditions progressively degrade Service member effectiveness. Fortunately, long-term remedies exist for slowing the rate of performance decline. These remedies, which must be introduced prior to combat, include safety, food intake, combat load, and physical fitness.

Safety

Safety, which encompasses such factors as using proper lifting techniques and staying alert and careful, is influenced by fatigue. Overtired Service members are more vulnerable to injury than those who are rested. After 72 hours of continuous combat, the tendency to seek shortcuts is very strong, and accident rates increase 50 percent. Fatigue affects all military systems, but it is especially hazardous when weapon systems are involved. Catastrophic accidents can occur when fatigued (and under-experienced) crews man weapon systems. Ways to safeguard Service members include developing and following safety standing operating procedures and increasing supervision during extended operations.

Food Intake

If Service members are too busy, stressed or tired to eat adequate rations during continuous operations, their caloric intake will be reduced. This may lead to both physical and mental fatigue and degraded performance. For example, in accidents judged to involve aviator fatigue, there is some indication that before the accidents occurred, the pilots had irregular eating schedules or missed one or more meals. In field tests done by the U.S. Army Natick Research, Development, and Evaluation Center, Natick,

MA, some personnel subsisting solely on meals, ready to eat (MREs) lost weight over just a few weeks. Leaders must encourage troops to eat all of the main items in their MREs, not just the candy, to get balanced nutrition. In various other field tests conducted during continuous operations, it was discovered that meals were frequently delivered late or missed altogether. Although the relationship between performance and nutrition is not clear, eating regularly is important in continuous operations, as well as providing hot meals at assigned times or when the workload has been sustained boosts morale.

Good nutrition is important. An inadequate diet degrades performance, reduces resistance to disease, and prolongs recuperation from illness and injury. When food and water become available in continuous operations, leaders must provide and maintain a supply of food that has the nutritional value commensurate with the physical activity and stress of battle. They must remind and encourage Service members to eat and drink properly. The excitement, stress, and rapid pace of events associated with field preparations can cause Service members to forget to drink liquids. Thus, they enter the early part of the field scenario inadequately hydrated. Dehydration may result, especially if the early scenario calls for assault of a position or rapid air/land deployment. Contributing to developing dehydration is the relative lack of moisture in MREs and other packets. In addition, Service members experiencing dehydration lose their appetite and reduce their food intake. This, in combination with dehydration, leads to degraded performance. Leaders must reemphasize drinking regimens to ensure that Service members are properly hydrated going into battle. Leaders must remind Service members to drink liquids in both hot and cold climates and must monitor fluid intake. If personnel drink only when thirsty, they will become dehydrated.

Combat Load

In combat, the load carried by a Service member significantly exceeds optimum recommended weights. In the case of a light infantry Service member, the combat load is as much as double the recommended load. Physical conditioning cannot compensate for this degree of excess. Service members tire faster and, in continuous combat, recovery from fatigue becomes more time-consuming.

When the Service member must carry excessive amounts of equipment, the effects of stress and lack of rest are magnified. Employing a load echelonment concept must be considered to ease the strain on Service members. In this concept, the unit separates an individual's equipment into two loads—fighting and existence. As the unit closes the objective, the heavier existence load is dropped and the Service member continues with the lighter fighting load.

Physical Fitness

Good physical conditioning delays fatigue, builds confidence, and shortens recovery times from illness and injury. It also prepares individuals to better cope with the physiological demands of stress. Service members in top physical condition can better control their internal physiological functions, which will improve their overall performance. Physical fitness—including aerobic fitness, muscular strength, and endurance—must be developed in all Service members to strengthen their ability to rebound from exhaustion. Aerobic fitness increases work capacity and the ability to withstand stress. While feelings of depression and moodiness accompany tiredness, aerobically fit Service members are affected less than those unfit. The ability to quickly recover from physically strenuous workloads is maintained by smart physical training, performed consistently and routinely. However, there is no evidence that good physical conditioning significantly reduces normal sleep requirements nor compensates for the deleterious impact of sleep deprivation on cognitive functioning. Sleep deprivation is discussed in greater detail in chapter 4.

2004. EFFECTIVE LEADERSHIP

The effective leader in combat is competent and reliable. He knows his job without question, and he can be counted on to do it regardless of the situation or circumstances.

Effective small-unit leadership reduces the impact of stress in several ways. Leaders understand the sources of combat stress and reactions to them. In addition, leaders manage stress problems to keep them from spreading throughout the organization by implementing the following actions:

- Continue mission performance; focus on immediate mission.
- Expect Service members to perform assigned duties.
- Remain calm, directive, and in control at all times.
- Let Service members know their reactions are normal and there is nothing seriously wrong with them.
- Keep Service members productive (when not resting) through recreational activities, equipment maintenance, and training to preserve perishable skills.
- Ensure Service members maintain good personal hygiene.
- Ensure Service members eat, drink, and sleep as soon as possible.
- Let the Service members express their thoughts. Do not ignore or make light of expressions of grief or worry. Give practical advice and put emotions into perspective.

A battalion commander in World War II made the following address to his Marines just before a new offensive:

> I know as well as you do that the going has been tough. Maybe I sweat it more than you do because I have more to worry about. But you've done a fine job, and I'm proud of every single man in my outfit. I assure you that everything possible will be done to give you the best available support, and I will not order you to attack unless I'm confident that you have a real chance to succeed . . . . The harder we fight now, the sooner we can finish this mess and get back to living the way we want.

That speech demonstrated a common sense approach in dealing with stress. Had this commander preached about the advantages of democracy over fascism or the evils of Hitler, his words would have fallen on deaf ears. Instead, he seized on the strong motivators of his troops: their *respect* for him as a leader, their desire to continue to be worthy of his respect, and their desire to share in the unit's accomplishments.

Generally, confidence is based on the technical and tactical competence exhibited by leaders and Service members. Since confidence is one of the strongest defenses against stress, its development is a major goal for every military leader. To gain confidence, Service members must believe in themselves, their equipment, other unit members, and their training. Above all, they must believe in their leaders' competence. Each of these beliefs is instilled and reinforced at every opportunity.

The following comments were made by a Marine colonel reflecting on his experience as a company commander in Vietnam.

> A feeling of helplessness will overtake men when they realize that they have been put in an untenable position. When the situation is so bad that the men cannot fight back, they are susceptible to fear. The

situation can be critical, but as long as the men can fight back, fear normally will not overtake them. The solution to this problem is: don't put your men in untenable positions. Granted, in every battle some men in a unit will find themselves in a situation where they can't fight back. This is inevitable especially in an attack. Proper use of fire team, squad, platoon, and company formations, as well as intelligent use of terrain, minimizes the danger of a large portion of the unit being pinned down to the extent where the men cannot return fire. If you are tactically and technically proficient, use common sense, apply the principles of war, and employ the firepower available to your unit, untenable situations can be avoided. Simply stated, know your job!

Those comments provide an insight and sensitivity to problems related to combat stress through small-unit leadership. His instincts told him how important it was to exhibit solid leadership qualities in order to allow his troops to operate at maximum effiiency. A unit builds confidence, esprit, integrity and cohesion when the *leaders know their jobs*.

2005. COMBAT STRESS MANAGEMENT BUILDING BLOCKS

Confidence in Leaders

Leaders must demonstrate effective leadership to earn their subordinates' loyalty and trust. Leaders are responsible for—

- Committing the unit to missions commensurate with abilities.
- Planning operations carefully and thoroughly.
- Preparing the unit to accomplish the mission.
- Leading and guiding the unit to mission accomplishment.

- Showing consistent good leadership that convinces subordinates their leaders know best what should be done, how it should be done, who should do it, and how long the task should take. Authority accompanies leadership beyond the automatic authority given by military rank and position. Authority and respect are earned based on confidence in a leader's ability to guide the unit to success.

Confidence in Training

Training helps Service members develop the skills required to do their jobs. Confidence is the result of knowing they have received the best possible training for combat, and are fully prepared. This confidence results from the following:

- Realistic training that ends with successful mastery.
- Relevance of training to survival and success on the integrated battlefield.
- Refresher and cross training.
- Systematic individual and collective training.

Confidence in Unit

Each Service member in a unit needs to become confident of the other unit members' competence. Individuals must stay and train together to gain that personal trust. Unless absolutely necessary, teams should not be disbanded or scrambled. Subunits in the same larger unit should have the same Standing Operating Procedures and training standards, so members can fit in quickly if teams have to be cross-leveled or reorganized after casualties occur. Confidence in the unit leads to feelings of security, which in turn allows members to sleep and positively focus stress. In combat, the unit must receive each member's highest commitment to unit loyalty. Mission accomplishment is the unit's highest priority.

Confidence in Equipment

Service members who learn to operate and maintain assigned equipment develop confidence in their ability to employ it. This, in combination with an individual's belief in his personal capabilities, raises overall confidence in fighting ability.

Cohesion and Morale

Good unit cohesion and morale offset the negative effects of combat stress. The foundation for any stress reduction program includes trust and confidence in the following:

- Fellow Service members.
- Competence and fairness of the unit leaders.
- Unit's technical abilities and military power.
- Equipment.
- Personal combat ability.
- Sense of support from the civilian community.
- Personal spiritual well-being.

Physical Conditioning

A strong relationship exists between physical stamina and the ability to resist combat stress. Good physical conditioning has physical and psychological benefits. Rigorous physical conditioning helps protect against the stress of continuous operations. A regular program of physical fitness to increase aerobic endurance, muscular strength, and flexibility is essential to combat readiness. As physical conditioning improves, Service members feel better about themselves, have greater confidence in each other, and their stress is reduced.

Unit training includes regular physical conditioning. This increases the members' tolerance to all types of stressors. The program is geared to the unit's combat mission, and exercises are tailored to the environment where the unit operates. The pace, length, and types of runs, road marches, and other activities are commensurate with the unit's need. Light infantry units need more demanding, longer road marches than maintenance units. Activities also include team athletics, which capitalize on cohesion-building aspects as well as physical benefits. The benefits of such a program include developing endurance through aerobic exercises, enhancing strength through weight training, and deprivation/physical stress training.

Family Care

Service members entering combat with financial worries or family problems risk breaking down under the additional stress of combat. Even positive but unfinished changes on the home front, such as a recent marriage or parenthood, can distract the Service members' focus on combat missions with worries that they will not live to fulfill their new responsibilities at home.

Leaders must be aware of this risk and assist members in handling personal matters before deployment. Pre-deployment and post-deployment family briefings should be conducted, and programs established to assist families before, during, and after deployment. When Service members know their families are cared for, they are better able to focus on their combat duties.

Coping with Individual Stress

Stress pushes the body to its limits and causes tension; relaxation reverses this process. Coping with personal stress is essential. Stress-coping skills are incorporated into unit training early, and command emphasis is placed on practicing them. Service members receive a block of instruction on stress-coping techniques, then are given supervised time each duty day to practice them quietly. After 3 or 4 weeks of practice, most Service members relax easily and quickly, even under highly stressful conditions. They will be able to naturally control stomach fluttering, heart rate, blood pressure, and stress.

The stress-coping exercises include deep breathing, muscle relaxation, and cognitive exercises. Deep breathing is the simplest to learn and practice; the others require longer instruction and more practice time. On request, the Combat Stress Control (CSC) team will provide instructional materials and assistance.

Deep-Breathing Exercise

This consists of slow, deep inhaling, which expands the chest and abdomen, holding it for 2 to 5 seconds, then exhaling slowly and completely, which pushes out the used air. This can be done for five breaths as a quick, mind-clearing exercise, or continuously to promote sleep. Abdominal or diaphragmatic breathing (making the stomach move the air, rather than the upper chest) is especially effective for stress control and, with practice, can be done simultaneously with task that require full attention.

Muscle Relaxation Exercises

These are more complex. They generally consist of concentrating on various muscle groups and the tensing and relaxing of limbs to relax the entire body. Quick versions for use in action consist of tensing all muscles simultaneously, holding for 15 seconds or more, then letting them relax and "shaking out the tension." Deep relaxation versions start in the feet and work up (or start in the head and work down), body part by body part (muscle group by muscle group), tensing and then relaxing each in turn, while noticing how each part feels warm after it relaxes.

Cognitive Exercises

These consist of self suggestion (positive self-talk); imagery (imagine being fully immersed in a deeply relaxing setting); rehearsal (imagine performing the stressful or critical task under pressure and doing it perfectly); and meditation (clearing the mind of all other thoughts by focusing on every breath and silently repeating a single word or phrase). These techniques involve creating positive mental images that reduce the effects of stressful surroundings, redirecting mental focus and learning to detach from stress. Service members are encouraged to practice stress management techniques and discuss their use in combat.

CAUTION

It is important not to use deep relaxation techniques at times when you need to be alert to dangers in your surroundings. Practice the quick relaxation techniques so you can use them automatically without distraction from the mission.

2006. CRITICAL EVENT DEBRIEF AND UNIT LEADER'S AFTER ACTION REVIEW

NOTE: On the subjects of Critical Event Debrief/After-Action Review, U.S. Army personnel will follow the guidance provided in FM 22-51, *Leaders' Manual for Combat Stress Control*, chapter 6.

Critical Event Debrief

A Critical Event Debrief (CED) is a structured group process designed to mitigate the impact of a critical event and to accelerate normal recovery of those personnel involved. The CED is normally conducted by a team composed of trained members, (medical officers, chaplains, mental health professionals, trained unit members). A CED's main value is to quickly restore unit cohesion and readiness to return to action, through clarifying what actually happened and clearing up harmful misperceptions and misunderstandings. It may also reduce the possibility of long term distress through sharing and acceptance of thoughts, feelings, and reactions related to the critical event. Ideally, CEDs are conducted 24 to 72 hours after the occurrence of the event, away from the scene and separate from any operational debriefing. The typical CED lasts from 2 to 3 hours. Situations that warrant a CED include:

- Death of unit member
- Death or suffering of noncombatants (especially women and/ or children)
- Handling of the dead, management of carnage or the sight of devastation during disaster work
- Serious friendly fire incident

- Situation involving a serious error, injustice or atrocity
- Terrorist attack.

After Action Review

Although a CED would be warranted following such traumatic events as those listed, if access to CED trained professional team is not available, small-unit leaders can modify their After Action Review (AAR) to assist in identifying the level of stress reaction of unit members. The AAR or "hotwash" is a routine practice used by leaders to debrief post-mission operations. Accordingly, it provides a familiar, non-threatening forum for leaders to identify the levels of stress reaction experienced by the unit members using the information in paragraph 1004. The leader's AAR may be sufficient to restore unit readiness for further action. It may also identify personnel who need immediate or later referral for chaplain or mental health/CSC help for combat stress-related symptoms. In those instances when access to CED trained facilitators is not available, the unit leader should incorporate the following questions in the AAR:

Fact Phase

Ask unit members to describe the event from their individual perspectives. What was each member's specific role in the event?

Thought Phase

What were each member's first thoughts at the scene (or when the incident was first heard of)?

Reaction Phase

What was the worst thing about the event? What was thought and/or felt?

Symptom Phase
Describe probable thinking, physical, and emotional responses both at the scene and a few days afterward.

Training Phase
Relay information regarding stress reactions and what can be done about them. If prepared handouts are available, distribute them. Include points of contact within the unit.

Wrap-up Phrase
Reaffirm positive things. Summarize. Be available and accessible. Debriefing team/unit leaders together decide which individuals need more help or referral.

2007. EFFECTIVE COMBAT STRESS CONTROL PROGRAM

An effective stress management program starts with early planning, continues during employment, and extends beyond the return home. The CSC program is comprised of three phases: predeployment, deployment and combat, and post combat.

Phase 1: Predeployment

During pre-deployment, some stressors facing Service members include long working hours, preparation for training, fear of the future, family worries, and anxiety about the unit's readiness. Signs of poor coping include insomnia, increased use of alcohol, marital problems, increased bickering in the unit, irritability, and suicidal feelings. The most important preparatory steps to take in the predeployment phase are to:

- Conduct unit training and mission rehearsals.
- Prepare for changed sleep schedules and jetlag.

- Attend to task assignments and allocations.
- Conduct equipment and supply maintenance checks.
- Attend to personal and family matters. (Call the Family Service Center.)
- Integrate new members into the unit positively and actively.
- Welcome "significant others" (not just entitled beneficiaries) in the Family Support network information tree.
- Brief as much information about the operation as possible, consistent with operations security measures.
- Actively familiarize the unit members with the stressors and horrors they may encounter (for example, watch videotapes from network television showing suffering civilians or atrocities), but always followed immediately with what the unit will do in those situations.
- Arrange for pre-deployment training and education, especially for refresher training of stress reduction techniques from local mental health professionals, a Special Psychiatric Rapid Intervention Team (SPRINT) or CSC team, if available.

Unit Training

Because unit leaders have combatted the stressors associated with garrison living and peacetime training, they have learned to know their Service members and what affects their performance. All Service members should believe unit training has prepared them thoroughly for combat. Stress is reduced if the training improves belief in themselves, and the unit's ability to fight and win. They are assured training is designed and developed to meet combat requirements, provides the ability to fight successfully, and that supporting units receive the same realistic training.

Realistic mission rehearsal helps desensitize Service members against potential combat stressors. For example, wearing and realistically training in protective gear is important. By doing so in pre-deployment training, Service members become less distressed in combat. The unit should be exposed to extensive drills wearing protective gear during live-fire exercises before employment. Such training has the two-fold, stress-reduction benefit of building confidence and preparing Service members for combat.

Leaders who provide their Service members with advance knowledge about the enemy prepare them for the stress created during deployment and initial enemy encounter. It is important during such training to talk realistically about enemy strengths and weaknesses as well as those of their own units. While inspirational pep talks are also important at this time, they should not include biased, inaccurate information. Leaders earn trust and respect if their troops perceive them as accurate, dependable sources of information.

Physical Fitness Training

In the time-urgent work of preparing for deployment, leaders assure that physical fitness is sustained, either by the work involved or by selective unit physical training exercises. New unit members who are insufficiently fit receive special training.

Stress-Coping Skills Training

During preparation for deployment, the leaders direct the unit to practice stress coping and relaxation techniques, and are positive role models by demonstrating use of these techniques. If necessary, the chaplain and mental health professional personnel available to the units can provide additional training.

Sleep Discipline

Before deployment, unit leaders must consider fatigue and sleep loss occurring during combat. The enforcement of work and rest schedules begins early in pre-deployment training. During continuous operations, fatigue caused by lack of sleep is a major source of stress. Breaks in combat are irregular, infrequent, and unscheduled. Extended sleep is unlikely. *Sleep logistics* is emphasized such that sleep and rest are allocated or supplied like rations, water, equipment, and ammunition. Sleep discipline training addresses the following points:

- A unit-specific work-rest-sleep plan is developed and practiced.
- The unit leader or commander is included in the allocation of sleep and rest time, as lack of sleep will impair his judgment and decision-making skills as much as those of his subordinates.
- The plan allows Service members at least 5 hours of uninterrupted sleep, ideally between 2400 and 0600, every 24 hours. Persons receiving only 5 hours per 24 hours over a period of several days will accumulate a significant sleep debt.
- Sleep priority is assigned to those whose judgment and decision-making are critical to mission accomplishment.
- If received frequently, 1 hour of sleep or even 15-minute naps help, but “slow mental starting” upon wake up can result.
- Relaxation exercises complement sleep schedules. These exercises are used as an alternative to regular sleep or as an aid to help Service members rest under difficult circumstances.

If preparation is made for jetlag, its effects are minimal. The ideal method is to seclude units up to several weeks before deployment, gradually changing work, eat, and rest cycles to coincide with the destination’s schedule. Although rapid deployment does not permit this method, there are ways to help make the transition to a new environment, such as changing to new sleep cycles collectively versus individually; sleeping up to 12 hours before operation start; ensuring leaders receive rest; and adjusting diet to meet energy needs. If the unit is divided into shifts, make sure that shift assignments are maintained from day to day, i.e., if a member is assigned to the 0600-1400 shift, keep him on that same shift each day.

Task Allocation and Management

Overloading Service members with tasks or responsibilities is another major source of stress. Allocating tasks fairly among available Service members improves unit effectiveness as well as decreases stress. Proper allocation of tasks include:

Selecting the Right Person for the Job

The right person is fitted to the right task according to the task requirement and the individual's talents, abilities, and training.

Duplicating Critical Tasks

Two Service members are assigned to a critical task requiring mental alertness and complete accuracy. They check each other's work by performing the same task independently.

Cross-Training

Each Service member is trained in a secondary duty position to ensure competently stepping into the position of another.

Developing Performance Supports

Develop standing operating procedures, checklists or other mental aids to simplify critical tasks during periods of low alertness.

Equipment Maintenance and Supply

During pre-deployment, the unit maintains its equipment and manages needed supplies. Once deployed for combat, Service members require confidence that supplies are ample and equipment is dependable. The following questions are important:

- Does the unit provide ample training in equipment maintenance and troubleshooting?
- Has the unit's equipment been field-tested under realistic conditions? For example, have Service members fired and cleaned their weapons while wearing full combat gear or protective clothing?
- Does the unit have sufficient ammunition, food, water, and other essential supplies?
- Does the unit have contingency plans for procuring and managing critical supplies if normal channels are disrupted?

Personal and Family Matters

Family stress adds to combat-imposed stress and causes distraction, interference with performance of essential duties, and a negative impact on stress-coping ability. This will result in the unit's inability to perform at peak. The unit should help the Service members resolve important family care matters before deployment and develop methods for helping families when Service members are deployed. Service members are encouraged to:

- Generate or update their wills.
- Finalize power of attorney for spouses.
- Update life insurance policies, including Servicemen's Group Life Insurance (SGLI).
- Ensure family automobiles are in good repair.

- Develop lists of telephone numbers of reliable points of contact for specific problems (mechanics, emergency transportation, babysitters, sources of emergency money, health care, etc.).
- Resolve major legal issues such as alimony payments, property settlements following divorces, and child support payments.

Role of Leaders

Unit commanders should:

- Brief families as a group before deployment or as soon as possible after deployment. Within the bounds of security limits, explain the mission's nature. Even if a mission is highly confidential, families benefit from such a meeting by being told of the support available to them while separated. They begin to solve problems and form support systems with other families. This includes an opportunity to discuss family questions and concerns. The Family Services Office, base mental health service or the chaplain's office assist in staging this briefing.
- Establish points of contact (e.g. Key Volunteer Network) to assist with family problems. These volunteers possess good working relations with the chaplain and mental health personnel to assist with the management of complex problems.
- Establish key volunteer communication and support networks. Commanding officers' spouses, or spouses of sergeants major are often good resources for developing and running such networks; however, the involvement of junior Service members' spouses is also crucial. Some of the most enthusiastic participants are tasked to make *outreach* visits and encourage shy or depressed spouses to participate.
- Have mental health professionals conduct meetings to discuss post-deployment problems. For example, some children have

difficulty adjusting to a parent's absence, and other family members may experience similar difficulties. Mental health professionals give families valuable information on these normal reactions and suggest ways to prepare for them.

Phase 2: Deployment and Combat

In addition to the normal stress associated with moving to a combat zone, Service members in this phase start worrying about their survival and performance under fire. Their thoughts become centered on fear of the unknown. Leaders should emphasize that stress under these circumstances and conditions is expected and is a natural reaction. This will help prevent "normal" stress reactions from escalating into extreme reactions.

Unit leaders should provide as much information as necessary to their survival and mission success, reinforce the stress control techniques learned during pre-deployment, and help their subordinates understand what happens to them when stressors occur.

Deployment Vehicle

The deployment vehicle—in most cases, an airplane—is a stressor by itself. If it is a commercial aircraft, in-flight problems are usually minor. However, if the unit deploys on a military aircraft, leaders should accomplish the following:

- Designate areas for light exercise and stretching to counter seating discomfort caused by a buildup of leg fluids.
- Ensure Service members drink enough fluids to prevent dehydration and have access to head/latrine.
- Adopt the activity schedule of the new time zone. If the unit is in the sleep cycle or is already in or about to enter the sleep

cycle, cover windows; reduce lighting; and issue earplugs, blankets, and pillows.

- Allow uninterrupted sleep. If a stopover occurs during a sleep cycle, do not waken Service members to eat or partake in activities. If the stopover occurs during an activity period, take full advantage of it by having Service members take washcloth baths, stretch, and perform head-and-shoulder rotations.
- Upon arrival in the area of operations, follow the schedule of the new time zone. Eat the next meal and go to bed on the new schedule. Doing so helps the Service members' bodies adjust.

Information Flow

Since uncertainty about the future is a major source of stress, timely and accurate information becomes vital. Lines of communications are clearly defined and kept open. Issuing warning, operation, and fragmentary orders is critical to ensuring adequate information flow. Informational meetings are conducted at regular intervals, even when there is no new information to disseminate. This reinforces the organizational structure and the importance of unit meetings as the source of current, accurate information. Reliable sources of information are especially important for countering rumors.

Service members also need information or *performance feedback* after mission completion. Merely engaging in a firefight or completing a mission is insufficient. Service members must be told how they performed as a group. The knowledge of mission accomplishment and progress builds unit cohesion, develops a winning attitude, and reduces the effects of stress.

Family Support

Marine Corps Family Team Building programs and corresponding programs for other Services provide family support throughout deployments. The Navy-Marine Corps Relief Society, Army Emergency Relief, American Red Cross, Family Service Centers, and other community agencies also provide direct assistance to family members. The Key Volunteer Network and the American Red Cross continue to function as conduits for emergency information between Service members and their families. Leaders need to educate Service members about these programs and agencies that are available to serve the needs of the community. Effective communication and caring support networks help to prevent anxiety while Service members are deployed and/or in combat.

Religious Ministry Support and Pastoral Care

The Religious Ministry Team is an integrated part of the battalion or unit. In the Marine Corps, a Religious Ministry Team (RMT) consists of a chaplain and Religious Program Specialist (RP) or chaplain's assistant. RMTs are organic to commands or are assigned by higher headquarters to provide direct religious ministry support to a command. (**NOTE:** In the Army, the same teams are called Unit Ministry Teams (UMT).) During operations, the RMT often travels with the forward battalion aid station (BAS) in order to minister to the wounded and to be with the Service members who are most likely to experience combat stress. Using their professional training, skills, knowledge, and relationship with the Service members, chaplains provide care focusing on prevention of mild and moderate combat stress reactions. The RMT's primary mission is to provide ministry and pastoral care to the troops, offering faith, assurance, and hope. Before, during, and after the mission, the team provides field services, sacraments, and counsel, which bring comfort, assurance, and encouragement to Service members as they integrate their experiences into their lives.

In addition to being a spiritual/pastoral mentor for Service members, most chaplains are trained in some form of CED process. Chaplains are effective CED team members as well as trainers of small-unit leaders (e.g., platoon leaders, noncommissioned officers (NCOs), corpsmen, etc.) in CED team member skills and stress management techniques. Appendix B explains the roles of the Religious Ministry Teams.

Physical and Recreational Activities

Service members need an outlet for the anger, frustration, hostility, and grief developed in combat. It is unwise to conduct sports and recreation activities under observation by the enemy, but these activities can be conducted further to the rear. Although makeshift athletic games help relieve stress, they should be conducted with caution. It is interesting to note that the most frequent medical problem in Operation Desert Shield/Desert Storm was sports injury. Other activities, such as listening to music, reading or practicing relaxation exercises are encouraged at rest and recuperation facilities. Time and access to personal hygiene items and facilities are psychologically valuable in combat.

Humanitarian assistance and community relations' projects provide excellent opportunities for Service members to constructively engage in meaningful work. Public Affairs Officers, Civil Affairs Officers, and chaplains are excellent personal resources for developing working relationships with local nongovernmental organizations (NGOs) and private voluntary organizations (PVOs) in the coordination of such projects.

Integration of New Unit Members

The arrival of replacements requires small-unit leaders to conduct continuous training programs. Leaders should be concerned with the various stressors affecting new Service members, as well as those Service members who have been around for a while, including seasoned combat veterans. Leaders teach newcomers to use stress coping skills. Replacements are quickly integrated into their units and become thoroughly acquainted with all aspects of the new unit. The unit also becomes thoroughly acquainted with the replacements. A brief orientation with a sincere welcome, with NCOs and officer support, eases replacement transition for combat.

It is important Service members adjust to a new unit quickly and effectively. If they do not feel they are part of the group, unit cohesion and morale suffer, decreasing unit effectiveness. New Service members in garrison are assigned sponsors to assist them, introduce them to the unit, and smooth their families' adjustment. In combat, replacements are placed with suitable groups of seasoned veterans. New unit members are much more likely to become battle or stress casualties than are members of a seasoned and cohesive group. The veterans need to give support and advice to the newcomers, by example and direct action, if necessary.

Phase 3: Post-Combat

Just as pre-deployment and combat are stressful, the period after combat is also difficult. Today's rapid transportation enables Service members to travel from the battlefield to their hometowns in 48 to 72 hours. This short time often does not give them reflection with their comrades. Units should therefore set aside time in the last few days before leaving the theater to conduct their own "End of Tour" debriefing in which they start at pre-deployment and talk about whatever stands out in their memories, good or bad, as they recount the operation up to its end. There should also be appropriate memorial ceremonies and rituals that formally bring the operation to a close. Awards, decorations, and other recognition must be allotted fairly by the commanders.

Unit officers, staff NCOs, and NCOs, assisted by the chaplains and mental health/CSC teams, prepare the Service members for problems encountered during family reunion. For example, most Service members expect to resume roles and responsibilities they had prior to separation. However, their spouses often resist giving up their new roles as decision makers and primary home managers. Also, a spouse may feel that his or her sacrifices during the Service member's absence have gone unrecognized. This feeling becomes an additional source of tension. If at all possible, the families should receive the same briefings or written materials. Families need to be reassured of their contribution. Key Volunteer Networks and other Marine Corps Family Team building programs, and corresponding organizations for other Services continue to help manage problems with reunion and adjustment.

Service members are briefed that startle reactions to sudden noise or movement, combat dreams and nightmares and occasional problems with sleeping, and feeling bored, frustrated and out of place are common when first returning from combat to a peacetime, civilian setting. The leaders, chaplains, and the CSC team emphasize the normalcy of such reactions. Service members are also advised on resources available to help deal with such symptoms, if they are persistent and become upsetting.

Chapter 3

Command Leadership Actions

3001. LEADERSHIP ACTIONS AND INTERVENTIONS FOR COMBAT STRESS

It is important for military leaders to know something about the treatment of combat stress reactions. As most cases of initial combat stress symptoms do not require medical treatment, it has been found that military leaders are often quite adept at treating less severe cases of combat stress. In most cases, debriefing the mission, including any traumatic events, coupled with rest, food, and sleep, is often sufficient to alleviate the symptoms. If the operational tempo permits the Service member to remain with his unit, and he responds to simple reassurance (e.g., "You just need rest, you'll be okay tomorrow"), he is not a casualty (by definition), and may not require further referral for specialized care. Military leaders in combat often make such decisions. When a Service member requires medical attention to rule out a possible serious physical cause for his symptoms, or because his inability to function endangers himself, the unit, and the mission, he should be transported to the BAS or equivalent nearest medical support facility. Interventions at the small-unit level include:

- If a Service member's behavior endangers the mission, himself, or others, the leader should take appropriate measures to control him.
- If a Service member is upset, let him talk about what is upsetting him, listen, then try to reassure him.
- If a Service member's reliability becomes questionable:

- Unload the Service member's weapon.
- Remove the weapon if there is a serious concern.
- Physically restrain the Service member only when safety is a concern or during transport.

- Reassure unit members that the signs are probably a normal combat stress reaction, and will quickly improve.
- If the combat stress reaction signs continue:
  - Get the Service member to a safer place.
  - Do not leave the Service member alone. Keep someone he knows with him.
  - Notify the senior NCO or officer.
  - Have the Service member examined by medical personel.
- If the tactical situation permits, give the Service member simple tasks to do when not sleeping, eating or resting.
- Assure the Service member that he will return to full duty as soon as possible.

The following quote was taken from an article about fear. It was written by a Marine who articulated his experiences during his tours in Vietnam. It provides intricate insight into how many Service members think and bond during combat.

> What should be done when men succumb to fear brought on by the realization that they are going to die? Before discussing this, first let us look at the men themselves. There are very few men who can be classified as cowards. Most men have too much self-respect to let their buddies down. It is the rare man indeed who will willingly violate the trust of his peers. The vast majority of men will give their lives rather than violate this trust. Proper training of combat troops, prudent leaders who are technically and tactically sound, and the reluctance of men to violate the trust of their peers are the foundation

of a solid combat unit. In this environment, courage and sacrifice are the rule, not the exception. At any given moment, anyone can be rendered ineffective by fear if one realizes that he is going to die. When this happens to a man, do not overreact. He knows he has let his buddies down, and he knows that his buddies and leaders know he has let them down. This is a very uncomfortable feeling for a combat Marine. The men around him will not make a big deal about his actions because they understand the situation. To make an issue of this situation will destroy the man and usually alienate the rest of your men. The man will most likely bounce back to his normal performance. The leader must keep in mind that today's coward is tomorrow's hero. If a man does not bounce back and continues to succumb to this type of fear, the Marines in his fire team will let you know when they have given up on him and no longer consider him trustworthy. Situations such as this are rare, so there is no need to make an example of the man. The biggest pitfall to be avoided is to judge the man's actions as a failure of your leadership as this will indeed cause you to overreact.

3002. SAFETY CONSIDERATIONS

Confiscation of a Service member's weapon should only be considered when it is clearly apparent that the Service member is unreliable and a safety hazard to himself or others. Service members identify with other Service members who are trusted and needed by their comrades. This is one of the strongest factors that aid in returning battle-fatigued members to effective duty. The small-unit leader relocates battle-fatigued Service members to a safer, quieter place to rest and work for several hours up to 1 or 2 days. If the unit cannot wait for the Service members to recover, it is necessary to evacuate them to the first level supporting medical unit. From there, every effort is made to move the Service members to a nonmedical unit or area (a tent or building of opportunity could suffice) for rest, replenishment, and reassurance. It should be made clear that the Service members are tired and in need of an opportunity to talk, to sleep, eat, and replenish fluids; they are not patients. Some small detachments without first sergeants or platoon/company organizations are attached to other units for support and rely on the effectiveness of small-unit leaders. Each Service member is accounted for and every effort is made to ensure strong lines of communications are in place and maintained between Service members and their original unit. Key to successful treatment is the return of the Service member to his original unit.

Actions to be taken for severely combat-stressed Service members are the same as those for the moderately combat-stressed, with one exception: medical personnel at the BAS level should evaluate severely combat-stressed Service members as soon as possible. Casualties will be treated and released within hours; held for rest and replenishment; or evacuated for further restoration. Service members who recover from combat stress reactions; return to their original units, i.e., same company or platoon; and are welcomed upon their return are less likely to suffer recurrence. Once rested and returned, they usually become healthy again. New joins and replacements in the unit are more susceptible to succumb to combat stress. Accordingly, this risk is reduced when Service members recovering from combat stress return to the same unit where their combat experience is known and welcomed. In rare instances, however, it is in the best interest of the individual to be reassigned to other jobs or units.

3003. COMBAT STRESS CASUALTY INTERVENTION MODELS

Management of combat stress reactions is unlike the treatment of physical trauma. Severely injured Service members are stabilized as rapidly as possible and then transferred to the rear. In all wars since World War I, combat stress casualties treated in the rear rarely return to full duty. In contrast, when the same casualties are treated near the front, approximately 75 percent return to full duty. Of those returned to full duty, only 10 percent experience continued symptoms requiring further treatment. Some studies suggest half of those treated at the rear go on to have chronic psychiatric symptoms, and approximately half return to full duty. Therefore, it is clearly in the Service member's best interest to be treated at the front and returned to duty. This also serves the best interest of the unit because it receives a veteran Service member back to duty, rather than breaking in a replacement.

Guidelines for dealing with Service members unable to function because of combat stress reaction are summarized in the memory aid **BICEPS.** BICEPS stands for brevity, immediacy, centrality (Marines)/command contact (Army), expectancy, proximity, and simplicity.

- **Brevity.** Critical Event Debriefing should take 2 to 3 hours. Initial rest and replenishment at medical CSC facilities should last no more than 3 or 4 days. Those requiring further treatment are moved to the next level of care. Since many require no further treatment, military commanders expect their Service members to return to duty rapidly.
- **Immediacy.** CSC should be done as soon as possible when operations permit. Intervention is provided as soon as symptoms appear.

- **Centrality (Marines)/Contact (Army).** Service members requiring observation or care beyond the unit level are evacuated to facilities in close proximity to, but separate from the medical or surgical patients at the BAS, surgical support company in a central location (Marines) or forward support/division support or area support medical companies (Army) nearest the service members' unit. It is best to send Service members who cannot continue their mission and require more extensive respite to a central facility other than a hospital, unless no other alternative is possible. The Service member must be encouraged to continue to think of himself as a warfighter, rather than a *patient* or a *sick person*. The chain of command remains directly involved in the Service member's recovery and return to duty. The CSC team coordinates with the unit's leaders to learn whether the overstressed individual was a good performer prior to the combat stress reaction, or whether he was always a marginal or problem performer whom the team would rather see replaced than returned. Whenever possible, representatives of the unit, or messages from the unit, tell the casualty that he is needed and wanted back. The CSC team coordinates with the unit leaders, through unit medical personnel or chaplains, any special advice on how to assure quick reintegration when the Service member returns to his unit.

- **Expectancy.** The individual is explicitly told that he is reacting normally to extreme stress and is expected to recover and return to full duty in a few hours or days. A military leader is extremely effective in this area of treatment. Of all the things said to a Service member suffering from combat stress, the words of his small-unit leader have the greatest impact due to the positive bonding process that occurs during combat. Simple statements from the small-unit leader to the Service member that he is reacting normally to combat stress and is expected back soon have positive impact. Small-unit leaders should tell Service members that their comrades need and expect them to return. When they do return, the unit treats them as every other Service member and expects them to perform well. Service members suffering and recovering from combat stress disorder are no more likely to become overloaded again than are those who have not yet been overloaded. In fact, they are less likely to become overloaded than inexperienced replacements.

- **Proximity.** In mobile war requiring rapid and frequent movement, treatment of many combat stress cases takes place at various battalion or regimental headquarters or logistical units, on light duty, rather than in medical units, whenever possible. This is a key factor and another area where the small-unit leader helps in the treatment. CSC and follow-up care for combat stress casualties are held as close as possible to and maintain close association with the member's unit, and are an integral part of the entire healing process. A visit from a member of the individual's unit during restoration is very effective in keeping a bond with the organization. A Service member suffering from combat stress reaction is having a crisis, and there are two basic elements to that crisis working in opposite directions. On the one hand, the Service member is driven by a strong desire to seek safety and to get out of an intolerable environment. On the other hand, the Service member does not want to let his comrades down. He wants to return to his unit. If a Service member starts to lose contact with his unit when he enters treatment, the impulse to get out of the war and return to safety takes over. He feels that he has failed his comrades and they have already rejected him as unworthy. The potential is for the Service member to become more and more emotionally invested in keeping his symptoms so he can stay in a safe environment. Much of this is done outside the Service member's conscious awareness, but the result is the same. The more out of touch the Service member is with his unit, the less likely he will recover. He is more likely to develop a chronic psychiatric illness and get evacuated from the war. This is one of the essential principles of CSC.

- **Simplicity.** Treatment is kept very simple. CSC is not therapy. Psychotherapy is not done. The goal is to rapidly restore the Service member's coping skills so that he functions and returns to duty again. Sleep, food, water, hygiene, encouragement, work details, and confidence-restoring talk are often all that is needed to restore a Service member to full operational readiness. This can be done in units in reserve positions, logistical units or at medical companies. Every effort is made to reinforce Service members' identity. They are required to wear their uniforms and to keep their helmets, equipment, chemical protective gear, and flak jackets with them. When possible, they are allowed to keep their weapons after the weapons have been cleared. They may serve on guard duty or as members of a standby quick reaction force.

If treated near their units, 65 to 85 percent of combat stress casualties treated in medical CSC facilities return to duty within 1 to 3 days. About 15 to 20 percent more return to duty in 1 to 2 weeks. Only 5 to 10 percent are sent home, and these usually have other problems in addition to combat stress reactions. If evacuated, few combat stressed Service members will return to duty. In fact, many are likely to be permanently disabled.

History shows that it is important to treat combat stress casualties as close to the front as practicable. Everyone is susceptible to the effects of combat stress, regardless of past performance, rank or duty. *Most Service members receiving combat stress control in accordance with the principles of BICEPS return to duty quickly.* The chances of later psychiatric problems are also reduced.

Combat stress reactions are inevitable in combat, but high stress casualties are not. History shows that highly trained and small cohesive units with good leadership have less than 1 such casualty for every 10 to 15 WIA, even in very heavy fighting. This is significantly fewer than the usual 1:3 to 1:5 ratio in high-intensity battle, and the 1:1 or 1:2 ratio in extreme fighting. Knowledge of factors that increase combat stress disorder in the tactical and overall situation encourages small-unit leaders, comrades, and the individual Service member to act towards building unit cohesion, sharing the burden, and reducing stress. The same measures, plus positive adherence to discipline and the law of war, also prevent combat stress-related misconduct that defeats the purpose of the mission. Tough, realistic training that builds confidence and caring within the unit in combat overcomes many adversities associated with combat stress in a combat environment. However, if a member fails to adequately respond to actions taken by the unit, then he should be evacuated for appropriate medical evaluation.

Chapter 4

Sleep Deprivation

4001. CHALLENGES OF SLEEP DEPRIVATION

People accumulate a "sleep debt" (cumulative loss of sleep over time) when they perform under limited sleep conditions. The only corrective measure for satisfying this sleep debt is sleep itself. Military operations, by their demanding nature, create situations where obtaining needed sleep will be difficult or impossible for more than short periods.

Continuous operations are military operations with many pulses of action every day and night, continuing for several days to weeks, which require careful planning and resource allocation to give everyone a minimum of 4 hours sleep in 24. (FM 22-51)

Sustained operations are continuous operations or combat with opportunity for less than 4 hours sleep per 24 hours for significant personnel, which may be brief or fragmented. (FM 22-51)

Accordingly, service members may have opportunities for only limited or fragmented sleep over an extended period. As a result of these periods of sleep loss, several combat tasks are likely to show decreased performance. These tasks include the following:

- Orientation with friendly and enemy forces (knowledge of the squad's location and maintaining camouflage, cover, and concealment).

- Coordination and information processing (coordinating firing with other vehicles and dismounted elements, reporting vehicle readiness, and communicating with the headquarters).
- Combat activity (firing from bounding vehicle, checking the condition of weapons, observing the terrain for enemy presence).
- Force preservation and regrouping (covering disengaging squads, marking the routes between locations, and conducting reconnaissance).
- Command and control activity (directing location repositioning, directing mounted defense, assigning fire zones and targets).

Continuous operations will potentially be more commonplace on the battlefield. In offensive operations, darkness is the time to retain or gain the initiative; while in defensive operations, obstacles can be employed with greater security during darkness. Forces can disengage undetected and threats to close air support lessen. The physical environment changes at night. As the air cools below ground temperature, inversions reduce visibility and hamper radar and radio signals. Conditions are optimal for using chemical weapons. Visual changes also occur. Without the aid of white light, there is no color perception. There is also a decrease in visual clarity, field of view, and depth perception. Targets take longer to engage. Preparation time increases two-fold to six-fold. Simple actions, such as the departure and return of patrols, become more complex and dangerous. Nighttime planning and coordination require greater attention. Navigation, adjusting fire, and munitions and/or target matching are more difficult. Precision is essential, but accuracy has a price. Service members tend to maintain accuracy at the sacrifice of speed. The adverse conditions associated with or generated by continuous ground combat at night will degrade the fighting performance of Service members, teams, and units. The almost complete mechanization of land combat forces and technological advances that permit effective movement at night, during poor weather conditions, and under conditions of limited visibility have largely overcome the reasons for "traditional" pauses in battle, such as darkness, resupply, and regrouping. New technologies have significantly increased the range, reduced the time, and changed the conditions over which battles are fought. For example, day/night-capable vehicles can operate for extended periods without re-supply, but they are limited by a crew's need to sleep. A Service member is not a machine and is, therefore, the weak link in the chain. The equipment can operate longer than the Service member who operates it, as the Service member must have sleep.

Commanders and leaders must ensure that all Service members obtain enough rest to counteract the effects of rapidly shifting from daytime to nighttime duty hours, or to extended work schedules. Implementing countermeasures that are designed to help Service members adapt to continuous operations conditions can satisfy this requirement. Neither leaders nor their subordinates can perform without rest or sleep. The Service member, the unit, and the leader are all affected by continuous operations. Generally at night, the cognitive and physiological resources of Service members are not at their peak, especially after a rapid shift from daytime to nighttime duty hours. Fatigue, fear, feelings of isolation, and loss of confidence may increase.

Non-stop, unrelieved combat operations (sustained operations) with little or no sleep degrade performance and erode mental abilities more rapidly than physical strength and endurance. Information gained from the Army Unit Resiliency Analysis Model shows that even healthy young Service members who eat and drink properly experience a 25 percent loss in mental performance for each successive 24-hour period without sleep. The mental parameters include decisionmaking, reasoning, memory tasks, and computational tasks. The loss may be greater for Service members who are older, less physically fit, or who do not eat and drink properly.

The effects of sustained operations are sometimes hidden and difficult to detect. Units are obviously impaired when Service members are killed or wounded in action or become noncombatant losses. They are further impaired when their troops are too tired to perform their tasks. Unlike individual performance, unit performance does not deteriorate gradually. Units fail catastrophically, with little warning.

A priority for fighting units is to assure that commanders and leaders are rested and able to think clearly. While this is obvious, it is a most difficult lesson for leaders to learn. During combat, commanders must focus on the human factor. They must assess and strengthen their units as they plan and fight battles. They must accurately decipher which units must lead, which must be replaced, where the effort must be reinforced, and where tenacity or audacity and subsequent success can be exploited. When leaders begin to fail, control and direction become ineffective, and the organization disintegrates. No fighting unit can endure when its primary objectives are no longer coordinated. Leaders must also prepare and precondition Service members to survive. It is particularly important that leaders conscientiously plan and implement effective sleep plans, because activities that are most dependent on reasoning, thinking, problem solving, and decision-making are those that suffer most when sleep and rest are neglected.

Some leaders wrongly believe that their round-the-clock presence during an operation is mandatory; they are unwilling to recognize that they, too, are subject to the effects of sleep deprivation. If the unit has been regularly trained according to the mission command philosophy, two benefits accrue. Not only will a leader be confident that in his absence his subordinates will adhere to his intent, but the trust he shows in his subordinates will continue to maintain unit morale and help ease some of the stress of the situation.

In future operations, the battlefield will become increasingly lethal. The threat of nuclear, biological, and chemical weapons will maximize confusion, uncertainty, and stress, which adversely impact our ability to move, shoot, communicate, and sustain. Sleep loss in this type of environment increases an already stressful situation.

4002. EFFECTS OF SUSTAINED OPERATIONS ON PERFORMANCE

A basic rule for continuous operations is planning ahead to avoid sustained operations, and provide members 5 to 6 hours sleep in 24. However, missions or enemy actions sometimes require exceptional exertion for several days with only unpredictable, fragmented sleep—as required in sustained operations. Sustained combat leads to exhaustion and reduction in effective task performance. Even during the first night of combat, normal sleeping habits and routines are abnormal. The Service member feels the effects of fatigue and the pressure of stress from noise, disrupted sleep time, and threat to life. While essential for endurance, sheer determination cannot offset the mounting effects of adverse conditions. Cognitive degradation involving poor decisionmaking begins during and after the first 24 hours of sleep deprivation.

Individual and unit military effectiveness is dependent upon initiative, motivation, physical strength, endurance, and the ability to think clearly, accurately, and quickly. The longer a Service member goes without sleep, the more his thinking slows and becomes confused. Lapses in attention occur, and speed is sacrificed to maintain accuracy. Continuous work declines more rapidly than intermittent work.

Tasks such as requesting fire, integrating range cards, establishing positions, and coordinating squad tactics become more difficult than well-practiced, routine physical tasks, such as loading magazines and marching. Without sleep, Service members can perform the simpler and/or clearer tasks—lifting, digging, and marching—longer than the more complicated or ambiguous tasks such as a fine hand-eye coordination sequence; i.e., tracking a target through a scope.

Sleep loss affects memory, reasoning, mental assessments, decision-making, problem-solving, subsequent actions, and overall effectiveness. While comprehension is accurate, reading speed slows and recall fails. For example, Service members may understand orders when reading them in documents, yet they are forgotten later when required. Individuals will forget or omit assigned tasks more often than they will make errors in carrying them out.

Leaders can expect declining moods, motivation, initiative, planning ability, and preventive maintenance. High motivation will only increase risk, due to impaired performance. Leaders must recognize erratic or unreliable task performance in subordinates, as well as in themselves. Alertness and performance decline gradually with partial sleep deprivation; that is, when sleep is limited to 4 to 5 hours per night. After 5 to 7 days of partial sleep deprivation, alertness and performance decline to the same low levels as those following 2 days of total sleep deprivation. After 48 to 72 hours without sleep, personnel become militarily ineffective.

Adverse Conditions

Continuous combat forces Service members to perform under adverse conditions that cause degradation in performance. Examples of adverse conditions follow.

Low Light Level

The amount of light available for seeing landmarks, targets, and maps is greatly reduced at twilight and night.

Limited Visibility

Smoke, fog, rain, snow, ice, and glare degrade a Service member's ability to see his environment and objects within it, as opposed to situations free of such conditions.

Disrupted Sleep Routines

People are accustomed to being awake or asleep during certain hours of the day or night. Disruption of the normal sleeping schedule causes degraded performance.

Physical Fatigue

Working the muscles faster than they can be supplied with oxygen and fuel rapidly creates "oxygen debt," eventually making these muscles unable to function until the deficits are made up during brief rests.

Sleep Loss

The muscles can continue to function adequately without sleep, but the brain cannot. Increasing sleep debt leads to subtle, but potentially critical, performance failures.

Sleep Loss Indicators

Indications of degraded performance symptoms become more prevalent as sleep debt accumulates. Performance is affected by the hours of wakefulness, tolerance to sleep loss, and the types of mental or physical work. Both mental and physical changes occur, with symptoms varying among individuals. Leaders must observe Service members for the following indications of sleep loss and degraded performance:

- Physical changes in appearance, including vacant stares, bloodshot eyes, pale skin, and poor personal hygiene. Other physical signs of sleep loss include the body swaying when standing, sudden dropping of the chin when sitting, occasional loss of hand-grip strength, walking into obstacles or ditches, low body temperature, slowed heart rate, and slurred speech.

- Mood changes, decreased willingness to work, and diminished performance go hand-in-hand. Service members may experience decreasing levels of energy, alertness, interest in their surroundings, and cheerfulness with a concurrent increase in irritability, negativity, and sleepiness. Some become depressed and apathetic. Others, for a time, can become energized by sleep loss, talk more, and may be more assertive without necessarily maintaining good judgment. Sleepiness and mood changes are not signs of weakness. After long periods of sleep loss, Service members go from being irritable and negative to dull and weary.

- Service members may feel more effort is needed to perform a physical task in the morning than in the afternoon. Exaggerated feelings of physical exertion may lead to work stoppage, especially between 0400 and 0700. During that time, the tendency to fall asleep is considerably more noticeable than other times.

- Both bickering and irritability increase with sleep loss. When Service members argue, it shows that they are still talking to each other and exchanging orders and messages. When arguments cease, especially after a period of increased bickering, Service members may be in a state of mental exhaustion.

- Comprehension and perception slow considerably. Individuals require extended time to understand oral, written or coded information; to find a location on a map and/or chart coordinates; to interpret changes in enemy fire patterns; and to make sense of things seen or heard, especially patterns. They may have difficulty with spot status or damage reports, and may be unable to assess simple tactical situations.

Loss of Concentration

Sleep deprivation causes the attention span to shorten. There is a loss of concentration on the job as dream-like thoughts cause lapses in attention. Leaders should watch for the following:

- Decreased vigilance. Personnel are less alert and fail to detect the appearance of targets, especially in monotonous environments. They may doze off at the wheel of moving vehicles.

- Distorted attention. Service members may imagine seeing things that are not there, e.g., "moving" bushes when in reality there is no such movement. The sleep-deprived brain can also misperceive bushes, rocks, people, vehicles or anything else and see them as something different, in very precise detail. Often the tired brain "sees" what it wishes were there (food, a bed); at other times, these illusions may be animals or other more bizarre things. But when the mind is alert for an enemy, the brain may generate a very convincing, detailed image of the enemy. Sometimes, but not usually, sounds or other sensations may accompany these illusions. They usually last only seconds, but can persist for minutes if not challenged, and rarely have even been "seen" by equally sleep-deprived comrades when told of them. It is essential for sleep-deprived unit members to check out any questionable things they see with their comrades, and to faithfully follow reporting and challenge procedures.

- Inability to concentrate; easily confused. Service members cannot keep their minds on what they are doing. They cannot follow multiple directions nor perform numerical calculations.

- Failure to complete routine tasks. Sleep loss interferes with completing routine individual tasks, such as drying the feet, changing socks or filling canteens when water is available. Tasks such as performing weapons checks may be skipped.

When a Service member cannot recall what he just saw, read, heard or was told by another individual, he is exhibiting a common sign of sleep loss. His memory loss is limited to recent events. For example, a sleep-deprived Service member may forget recent target data elements or recall them incorrectly and have difficulty learning new information.

4003. ACHIEVING SLEEP IN COMBAT

Sleep deprivation produces stress and, therefore, sleep management is important. Sleep management is a combat multiplier. Planned sleep routines are important for keeping the unit, the individual Service members, and the leader himself functioning as required while reducing sleepiness during continuous combat. Since leaders are responsible for planning sleep routines, they need a basic understanding of the physiological and behavioral aspects of sleep and their impact on performance. The following paragraphs provide this information.

Rhythmic Variations

There are rhythmic variations in individual performance based on a predictable physiological and behavioral cycle that comprises about 24 hours. The 24-hour, day-night/work-rest cycle is called the *circadian rhythm*. Because traveling across a half-dozen time zones disrupts the usual relationship in the day-night/work-rest cycle, for a few days Service members are not sleepiest at their usual sleep period of 2400 to 0600, new-locale time. Allowing sleep about 1200 to 1800, new-locale time, will only delay their adaptation to their new locale. Leaders must instruct troops to go to bed between 2400 and 0600 new-local time to establish a new circadian rhythm.

Another example of circadian rhythm is body temperature. Although one's "normal" temperature is 98.6 degrees, this is really an average or midpoint of a daily swing from 96.8 to 100.8 degrees. For someone accustomed to working days and sleeping nights, body temperature would fluctuate approximately as indicated. There is a well-established link between body temperature and sleepiness and/or performance slumps. Performance parallels body temperature. The higher the body temperature, the better the performance. As body temperature decreases, mood and motivation decline with a concurrent increase in sleepiness and fatigue.

Impact upon performance is most pronounced during the *circadian lull*, which is roughly 0200 to 0600 hours. During this time, performance declines about 10 to 15 percent. In sleep-deprived Service members, this decline may reach 35 to 40 percent. If the day-night/work-rest cycle is disrupted, performance suffers because the Service member is sleepy during the new work period and awake during the new sleep period. The body needs several days to adjust to the new schedule. Critical hours for sleep are between 0200 and 0600 when *anchor sleep* (the most beneficial sleep) is taken. The body is at its lowest temperature during this period. This is the best time for sleeping, but not for napping. To prevent sleep inertia, naps should always be taken at times other than the lowest point in body temperature.

Leaders need to calculate the difference in time zones and make the necessary schedule changes. Leaders will need day-and night-fighting teams. Members acclimated to working days and sleeping nights should be scheduled to work nights and sleep days.

Their performance slump/optimal time to sleep would be 2400 to 0600, new-locale time. Deployment, pre-combat, and combat are not usual circumstances. If certain Service members must have an offset circadian timing from the rest of the unit, a special effort must be made to establish their sleeping time. Obviously, troops must sleep whenever possible. If a planned sleep schedule cannot be followed, however, performance is enhanced if sleep coincides with the low point in body temperature.

Adjusting to new circadian rhythms is a slow process, taking 3 to 6 days to come "in phase" with a new schedule. Leaders should devise a sleep schedule that provides for sleep at the same time of day or night every 24 hours. Sleep schedules that provide for sleep at different times of day or night are less valuable and are detrimental to quality sleep and optimal performance.

Sleep Shifts

Staggered work schedules can be set up for two shifts working 4 hours on/4 hours off, 6 hours on/6 hours off, and 12 hours on/12 hours off. See Table 4-1. Each shift follows the same schedule daily. It is better to maintain regular shift schedules than schedules that continually change.

Sleep/Rest Guidelines

Leaders should use the following sleep and/or rest guidelines in this section to enhance individual and the unit performance in continuous operations.

- Know personal tolerance for sleep loss and those under your command; major individual differences are not easily changed. Individuals who are unable to sleep during pre-deployment and deployment stages should be encouraged to practice relaxation exercises (see paragraph 2005).

Table 4-1. Sleep Shifts.

4 HOURS ON/4 HOURS OFF						
Shift	2400-0400	0400-0800	0800-1200	1200-1600	1600-2000	2000-2400
1	SLEEP	DUTY	SLEEP	DUTY	SLEEP	DUTY
2	DUTY	SLEEP	DUTY	SLEEP	DUTY	SLEEP

6 HOURS ON/6 HOURS OFF				
Shift	2400-0600	0600-1200	1200-1800	1800-2400
1	SLEEP	DUTY	SLEEP	DUTY
2	DUTY	SLEEP	DUTY	SLEEP

12 HOURS ON/ 12 HOURS OFF		
Shift	2400-1200	1200-2400
1	SLEEP	DUTY
2	DUTY	SLEEP

- Ensure that Service members fully use their breaks and other opportunities for rest. Encourage them to waste no time in getting to sleep. Undisturbed, prolonged sleep is the most desirable use of rest opportunities. When there has been sleep loss but little physical exertion (e.g., manning communications, operating a radio), mild physical exercise such as walking around when conditions permit, can help maintain alertness.

- Encourage Service members to sleep, not just rest, by creating the most conducive environment possible for sleep: quiet, without interruptions (or earplugs); dimness or darkness (or with eye cover); not overly warm or cold.

- Do not allow personnel to sleep in unsafe conditions. Enforce strict rules designating sleep areas and requiring perimeter guards. Require day and night guides for all vehicles to prevent Service members from being accidentally run over.
- Ensure that Service members follow sleep schedules or routines. The field commander who does not enforce a sleep schedule or routine leads his troops into an environment that increases the opportunity for hazardous conditions to be encountered while in continuous combat. Taking naps is not a sign of low fighting spirit or weakness; it is a sign of foresight.

Measuring Sleep Loss

Sleep loss can be measured by:

- Keeping a sleep and/or activity log. From pre-deployment to post-deployment, log sleep and nap periods. Service members need 4 to 5 hours per 24-hour period; 6 or 7 hours is optimum. If they receive less, the first chance for a long rest period must be used for sleep.
- Observing performance and asking questions. Look for the indications of sleep loss—such as increase in error occurrence, irritability, difficulty understanding information, and attention lapses—with concurrent decreases in initiative, short-term memory, and attention to personal hygiene. Confirm sleep loss by asking the obvious question: "When did you sleep last and how long did you sleep?"

Sleep Loss Alternatives

Ways to overcome performance degradation include:

- Upon signs of diminished performance, find time for members to nap, change routines or rotate jobs (if cross-trained).

- Have the Service members most affected by sleep loss execute a self-paced task.
- Have Service members execute a task as a team, using the buddy system.
- Do not allow Service members to be awakened for meals while in flight to a new location, especially if the time zone of the destination is several hours different than that of point of departure.
- Insist that Service members empty their bladder before going to bed. Awakening to urinate interrupts sleep, and getting in and out of bed may disturb others and interrupt their sleep.
- Allocate sleep by priority. Leaders, on whose decisions mission success and unit survival depend, must get the highest priority and largest allocation of sleep. Second priority is given to Service members that have guard duty and to those whose jobs require them to perform calculations, make judgments, sustain attention, evaluate information, and perform tasks that require a degree of precision and alertness.

4004. SLEEP/REST PLANNING

Sleep/rest planning applies to the pre-deployment, deployment, pre-combat, combat, and post-combat stages of battle.

Pre-Deployment Stage

Using mission-scenario operation guidelines, determine periods available for sleep and the total number of sleep hours possible. Because continuous operations requirements may change, alternate sleep routines should be planned. Become familiar with the area where the combat unit will sleep; For example, some may have to sleep in mission-oriented protective posture (MOPP) IV. If sleeping in MOPP IV is anticipated in combat, practice it during the pre-deployment stage. Prior experience reduces stress, so practice anticipated sleep routines before continuous operations.

Deployment Stage

Since sleep will be reduced during deployment, follow preplanned sleep routines. The prudent commander will choose a 4-hour on/4-hour off, 6-hour on/6-hour off, or 12-hour on/12-hour off shifts from the start. Take into account that Service members on night duty will need to sleep during the daytime. Provide night-shift personnel with separate sleeping quarters to avoid disruption of their sleep period.

Pre-Combat Stage

In general, people are most effective during the afternoon and are least effective just before dawn. Without prior adjustment to the new time zone, which naturally occurs in 3 to 5 days, leaders can expect degraded daytime performance. The reason is that 0200 to 0600 hours home-base time is the low point in performance efficiency and should be considered when planning workloads.

Combat Stage

Every effort should be made to avoid situations where all personnel are physically and mentally exhausted simultaneously. Make the most of any lull during the combat phase by sleeping briefly. Complete recovery from sleep loss may not be possible during intense combat, but limited sleep is helpful. Uninterrupted short sleeps of 15 minutes or longer are beneficial to partially recovering alertness. Sleep during the combat stage may be risky, however, because a Service member may wake up feeling groggy, confused, sluggish, and uncoordinated. It may take his brain from several seconds to 15 minutes to "warm up." Individuals differ in how quickly they take to wake up, but it tends to be worse when the body expected to go into deep sleep, and to get worse with increasing sleep loss. Activities that increase circulation of warm blood to the brain, like moderate exercise or drinking a hot beverage, may shorten the start-up time.

Post-Combat Stage

It is important to make up sleep debt, but experts disagree about the amount of recovery time needed. Some say the hours of sleep needed for recovery after sleep deprivation are less than the amount lost. It is well known and documented that lost sleep is not made up hour-for-hour. Most experts agree that immediately following continuous combat, Service members should be allowed to sleep up to 10 hours. Longer sleep periods are not desirable because they cause "sleep drunkenness" and delay in getting back to a normal schedule. After the first sleep period of up to 10 hours, Service members should return to the regular sleep routine. Sleep inertia lasting longer than 5 to 15 minutes and increased sleepiness may occur for as long as a week following sustained combat. Some experts recommend that 4 of the first 8 hours of recovery sleep should be at the 0200 to 0600 sleep time, and they suggest the following guidelines for complete recovery from the effects of sleep loss:

- 12 hours for sleep and rest after 36 to 48 hours of complete sleep loss with light to moderate work load (fatigue may linger for 3 days).
- 24 hours for sleep and rest after 36 to 48 hours of sleep loss with high workload (12 to 16 hours per day).

- 2 to 3 days time off after 72 hours or more of acute sleep loss.
- As much as 5 days for sleep and rest following 96 hours or more of complete sleep loss.

Most experts agree that 10 hours of sleep is the maximum needed, with the additional 2 hours used for rest. It is doubtful that a Service member could continue past 72 hours of wakefulness. Should this occur, a couple of nights with 10 hours of sleep are more beneficial than an excess of 10 hours during one sleep period. If Service members have not slept for 36 to 48 hours or more, they should avoid sleep of less than 2 hours, especially between 0400 and 0600. A too-short sleep period at the wrong time may cause a long period of sleep inertia. After 96 hours of total wakefulness, 4 hours of sleep may provide substantial recovery for the simpler, less-vulnerable tasks. Recovery continues with additional days of 4 hours of sleep per 24 hours. Complex leadership tasks may require longer recovery sleep, but sleep until fully satisfied is not necessary.

Sleep loss alone does not cause permanent health problems, nor does it cause mentally healthy people to become mentally ill. Reduced sleep (from 8 to 4 hours) does not cause physical harm. Hallucinations may occur, but they disappear after recovery sleep. Clinical laboratory tests show that total sleep loss of over a week does not pose serious health problems. It is doubtful that Service members could stay awake for such an extended period, and it is not suggested that Service members try to endure long periods without rest. However, the effects of sleep loss, such as inattentiveness and poor judgment, may be harmful (such as falling asleep at the wheel of a vehicle).

Sleep cannot be stored in our bodies for emergency use. Sleep of more than 7 to 8 hours before deployment does not "store up" excess sleep, but sleep taken immediately before a deployment can prolong activity. Therefore, it is important to begin continuous operations in a rested state. During daytime or early morning naps, many Service members experience vivid dreams as they fall asleep and often wake up frightened. Leaders should inform their troops that this occurrence is both common and normal during daytime sleep. If a single, unbroken period of 4 to 5 hours is not available for sleep, "power naps" of 15 to 30 minutes, although less recuperative, can be taken. Leaders must capitalize on every opportunity for a "power nap." Merely resting by stretching out does not take the place of sleep. Only sleep can satisfy the need for sleep.

Appendix A

Suicide Awareness Information

SIGNS OF SUICIDE

A person contemplating suicide—

- Believes he or she is in a hopeless situation.
- Appears depressed, sad, tearful; may have changes in patterns of sleep and/or appetite.
- May talk about or actually threaten suicide, or may talk about death and dying in a way that strikes the listener as odd.
- May show changes in behavior, appearance or mood.
- May increase or start drug or alcohol use.
- May injure self or engage in risky behavior.
- Abandons planning for the future.
- May start withdrawing from others, including family and close friends.
- May give away possessions.
- May appear apathetic, unmotivated, indifferent.

PREVENTION KEYS

The small-unit leader may be the first to identify the "signals" of a potential suicide-risk Service member. The keys to prevention are to provide aid to persons at suspected risk and follow the acronym AID LIFE.

A	Ask: "Are you thinking about hurting yourself?"
I	Intervene immediately.
D	Do not keep a secret concerning a person at risk.
L	Locate help (NCO, chaplain, corpsman, doctor, nurse, friend, family, crisis line, hospital emergency room).
I	Inform your chain of command of the situation.
F	Find someone to say with the person. Do not leave the person alone.
E	Expedite! Get help immediately. A suicidal person needs the immediate attention of helpers.

Appendix B

Religious Ministry Team's Role

(Adapted from Army Field Manual (FM) 22-51, *A Leader's Manual for Combat Stress Control.*)

This appendix addresses the general role of the ministry team in the commander's program of combat stress control and in battle fatigue ministry. In the Marine Corps ministry teams are called Religious Ministry Teams (RMTs) and consist of chaplains and Religious Program Specialists (RPs) or chaplain's assistants (In the Army they are called Unit Ministry Teams). The RMT is assigned to a command or designated by higher headquarters to be responsible for the direct religious ministry support to the command. RMTs provide professional ministry support to leaders in fulfilling their combat stress identification and intervention responsibilities. RMTs can also assist in training leaders to recognize combat stress symptoms.

EFFECTS OF STRESS

Negative effects of stress can be lessened when members are prepared physically, emotionally, and spiritually prior to combat/operations. RMTs can assist in preparing Service members to manage combat stress with training before and during deployment. This training helps Service members draw upon their personal religious faith, spiritual strength, and values as well as to share strength and confidence during any operation.

RELIGIOUS MINISTRY TEAM RELATIONSHIP

The RMT's established relationship with the command promotes trust with the troops. As integral members of their commands, RMTs are trained and ready to respond to the needs of Service members experiencing combat stress. A person-oriented resource, the chaplain and RP team minister to combat stress casualties, especially Service members having less severe difficulties who have rapid replenishment potential.

SPIRITUAL VALUES

Service members' inner resources are often based on their personal faith, religious beliefs and spiritual values. In combat, Service members may show more interest in their religious beliefs. When religious and spiritual values are challenged during the chaos of combat, Service members may lose sight of inner resources that sustain them. Service members can then become targets of fear, despair, hopelessness, and eventually, combat stress casualties. They are also at risk for committing misconduct stress behaviors. Chaplains are a source of direction and stability to Service members experiencing these dilemmas and seeking to refocus their personal beliefs and spiritual values.

RELIGIOUS MINISTRY TEAM SUPPORT

Ministry teams can provide preventive, immediate, and replenishing spiritual and emotional support and care to Service members experiencing combat stress.

Preventive

The RMT's religious ministry support assists in preventing combat and misconduct stress behaviors through establishment of a presence within the unit. It is important for the RMT to be present with Service members when the unit trains and when it deploys. The chaplain and RP can be a calming influence on members and can help Service members strengthen or regain values important to them. Chaplains can help prevent combat stress and misconduct by:

- Being present with the Service members and deploying with the unit.
- Providing opportunities for private and group prayer and worship.
- Supplying personal religious articles and materials.
- Reading the scriptures with Service members.
- Providing sacraments as the situation allows.
- Counseling Service members, allowing them to work through stress, fear, anxiety, anger, and frustration.
- Visiting Service members in work and living areas.
- Assisting Service members and families prior to deployment with preparation for geographical separation and an uncertain future through Marine Corps Family Team Building (MCFTB) programs that emphasize family strengths. This helps Service members to know that their families are cared for during deployment.

Immediate

RMTs can assist commanders in the identification of Service members experiencing combat stress. Chaplains often work closely with the medical officer and are trained to recognize the signs of combat stress and provide religious support to Service members experiencing combat stress. Chaplains assess spiritual needs and provide the appropriate religious ministry. This may include such things as:

- Presence with the Service member.
- Conversation and counseling, providing opportunities to share fears, hopes, and other thoughts.
- General prayers; prayers for the individual or fallen comrades.
- Rites, sacraments, and ordinances as appropriate.
- Reading from scriptures.

Replenishing

Following an engagement, the unit may require reconstitution through the addition of new personnel. The RMT can assist surviving Service members rebuild their emotional, psychological, and spiritual strength. During this time, RMTs continue their ongoing direct religious ministry that includes:

- Coordinating the availability of worship services, sacraments, rites, and services and/or ceremonies honoring the dead.
- Assisting with the integration of personnel replacements.
- Enabling the grief process through personal counseling and memorial services.

- Reinforcing the Service member's sense of self-worth and hope.
- Assisting with CEDs to provide opportunities for Service members to talk about their combat experiences and to facilitate integration of the combat experience into their lives.
- Providing leadership training and supervision of CEDs.
- Providing personal religious articles and materials.
- Providing or arranging faith group coverage in the command or for other units.
- A chaplain's ability to relate the spiritual aspects of life to the Service member's situation is an essential element in the replenishment process. Religious ministry can assist Service members in achieving emotional and spiritual wholeness.

Reintegration

After deployments and operations, RMTs can assist reintegration of the individual Service member into family relationships and society at large. Many religious ministry support programs remain the same. Expanded ministry may also include providing the following:

- Worship events for the entire unit.
- Worship events for varied faith groups.
- Briefings that help Service members recognize, prepare for, and master the stressors of reunion with family.
- Structured events to assist Service members returning to family and civilian life.
- Opportunities for Service members to experience and understand the forgiving and unchanging love of God.

Appendix C

Glossary

Section I. Acronyms

AAR	after action review
BAS	battalion aid station
BICEPS	brevity, immediacy, centrality, expectancy, proximity, and simplicity
CED	critical event debrief
CSC	combat stress control
MCFTB	Marine Corps Family Team Building
MCRP	Marine Corps Reference Publication
MOPP	mission-oriented protective posture
MRE	meals, ready to eat
NCO	noncommissioned officer
NGO	nongovernmental organization
PVO	private voluntary organization
RMT	religious ministry team
RP	religious program specialist
SGLI	Servicemen's Group Life Insurance
SPRINT	Special Psychiatric Rapid Intervention Team
UMT	unit ministry team
WIA	wounded in action

Section II. Definitions

circadian rhythm—Rhythmic variations in individuals; predictable physiological and behavioral cycle of about 24 hours.

combat stress—The mental, emotional, or physical tension, strain, or distress resulting from exposure to combat-related conditions. Combat stress is the net result of many stressors (e.g., fear of death, fear of failure, other intensive painful emotions [like grief], uncertainty, boredom, worry about what is happening back home, and the many physical and mental demands of combat duties) at any given time.

combat stress behaviors—The generic term that covers the entire range, from highly positive to totally negative behaviors which can be elicited by combat stress.

combat stress reaction—The common, predictable, negative, emotional and physical reactions of normal people to the "abnormally high" stressors of combat, which by definition interfere with mission performance or well-being, but which can be treated according to the BICEPS principles, with reassurance of normality and recovery, with rest, replenishment, and activities to restore confidence. The combat stress reactions do not include the misconduct stress behaviors, since the latter require disciplinary action.

continuous operations—military operations with many pulses of action every day and night, continuing for several days to weeks, which require careful planning and resource allocation to give everyone a minimum of 4 hours sleep in 24.

stress—An internal process of an individual for preparation in dealing with, or reacting to stressors. Stress involves physiological stress reflexes, which ready the body for fight or flight, such as increased release of adrenaline, heart rate, and blood flow to brain and muscles, with subsequent increase in alertness. Stress also involves the accompanying emotions (stress responses) such as anxiety, anger, or guilt. Stress may or may not involve conscious awareness of the threat, but the stressor must be perceived at some level of awareness. The amount of stress experienced depends on the appraisal of the stressor and its context by the individual, even if that perception is incorrect. The stress process includes psychological defenses which may shield the individual from perceiving more threat than he can tolerate.

stress behaviors—Stress behaviors are the observable behaviors that an individual exhibits as a result of internal stress. They may be intended to overcome and turn off the stressor, escape it, adapt to an inescapable stressor, or make the uncomfortable stress symptoms more tolerable. A stress behavior may simply be a side effect of the tension, which is part of stress. Any of these different types of stress reactions may be successful, unsuccessful, adaptive, or maladaptive. They may make the situation better, worse, or not influence it at all. They may resolve one stressor but create new ones.

stressor—Any event or situation that requires a non-routine adaptation or behavior change, creates conflicting motives within the individual, or is perceived by the individual as a threat to his well-being or self-esteem. Combat stressors are any that occur in the context of performing the combat mission (whether under fire or not).

sustained operations—Continuous operations or combat with opportunity for less than 4 hours sleep per 24 hours for significant personnel, which may be brief or fragmented.

Appendix D

References

DoD Directive

6490.5, *Combat Stress Control (CSC) Programs*

Army Field Manual (FM)

8-51, *Combat Stress Control in a Theater of Operations, Tactics, Techniques, and Procedures*

22-9, *Soldier Performance in Continuous Operations*

22-51, *Leaders' Manual for Combat Stress Control*

22-100, Army *Leadership*

90-22, *(Night) Multiservice Night and Adverse Weather Combat Operations*

Miscellaneous

U.S. Army Aeromedical Research Laboratory and U.S. Army Safety Center, *Leader's Guide to Crew Endurance*

http://www.mcu.usmc.mil/TbsNew/Coverpag.htm, *Human Factors in Combat: Topics in Combat Leadership that Help Preserve the Fighting Spirit and Prevent Psychiatric Casualties*

Made in United States
North Haven, CT
04 November 2026

21774378R00134